全球先锋设计系列

设计师国际名片创意作品

善本 编

中国林业出版社

图书在版编目（CIP）数据

黑：11位国际名师的巅峰力作：英汉对照 / 燃燃著.
-- 北京：中国林业出版社, 2012.7
　ISBN 978-7-5038-6571-8

Ⅰ.①黑… Ⅱ.①燃… Ⅲ.①建筑设计－世界－现代－图集 Ⅳ.①TU206

中国版本图书馆 CIP 数据核字 (2012) 第 096407 号

中国林业出版社·建筑与家居出版中心
责任编辑：成海沛 纪 亮
书籍设计：陈志平
策划专访：5只猫
英文翻译：那晓丹
在线对话：315435059（QQ）

出版：中国林业出版社
　　（100009 北京西城区德内大街刘海胡同 7 号）
网址：www.cfph.com.cn
E-mail：cfphz@public.bta.net.cn
电话：（010）8322 8906
发行：新华书店
印刷：恒美印务（广州）有限公司
版次：2012 年 7 月第 1 版
印次：2012 年 7 月第 1 次
开本：1/16
印张：16
字数：200 千字
定价：128.00 元

购买本书凭密码增送高清电子书

密码索取方式
QQ：179867195　　E-mail：frontlinebook@126.com

BLACK

黑

文 / 燃燃

黑色，纯净、永恒，品质至臻，具有不可超越的虚幻和无可比拟的高贵，是整个色彩世界的主宰。近墨者黑，水墨散淡悠远的意境，是中国独有的文化精髓。

冷冰川的作品中，黑色是如此的绚丽多姿：春风涤荡的夜晚，月色如甘洌的泉水，连绵不绝的远山，古老恬然的村庄，茂盛低矮的小树，待收割的一望无际的稻田、性感妖媚的女人，所有的一切，在黑色之中，私藏着不动声色的惊艳，散发着若隐若现的芬芳。

魏琳的《月夜》油画中，黑色令人迷醉：天空中飘洒着细密的小雪，雪落无声。在水银样流动的月光里，羊群行走，湛蓝色的湖水泛起微光。无垠的雪地，细碎而神秘的熠熠闪亮。

顾城在《静静的落马者》中，关于黑色的诗句：
那些叶子上缠绕的黑夜的茵丝，
你无疑会从这里开始，
像沙子、像鱼、像白衣少年的奇迹，
你热切地等着，你将孤身前往。

窦唯在《黑梦》中对黑色的吟唱：
到处寻找寻找安慰，
对我来说那太珍贵，
黑色梦中我去安睡，
梦中没有错与对。

艾未未在《为何建筑师爱穿黑》中对黑色的态度：
消失，消失在世间。

韩国设计师玄源明在《纯净的墨黑》中对黑色的描绘：这里所有的元素都是纯墨色的，不掺杂任何其他颜色。家具和餐具都采用漆黑的镀锡铁皮作为材料，用印度墨喷在韩国纸上，并向前抛掷，形成浑然的黑色。

……

黑色，是纯粹的精神自由；是对设计的探索前行；是丰富的安静；是心中期许已久且尚未到来的美好；是全球先锋的设计概念；是抽象；是不确定；是风格；是锐度；是尖利的棱角；是深不可测的锋芒；是让人倍感惊喜的意外；是没有预约的真实来临；是繁盛的花朵；是万物厮杀，胜者为王；是想像力量的无限发挥；是创造潜质的无限延展；是未来生活方式的无限可能。

燃 燃

5只猫主编

ICIAD 深圳秘书长

曾任《现代装饰》主编；

2009年，出版《设计 灵魂的舞者》（上、下册），书中收录香港设计名师靳埭强、梁志天、张智强、林伟而、刘小康、方振华、Marc brulhart、陈德坚、何宗宪和80后设计师的精彩对话；

2009年，5只猫中国室内设计新闻网（www.5cats.cc）主编；

2011年，国际室内建筑师和设计师理事会(ICIAD)深圳秘书长。

目录 CONTENT

007 KENZO TAKADA   Dream Warms All
高田贤三   梦想给每个人温暖

019 MASAHIRO HARADA   The Age of Architecture=Art Is Coming
原田真宏   建筑＝艺术的时代即将到来

045 MI5   All about Nature
MI5   任何的创作都是回归自然

083 WON-MYOUNG HYUN   Pioneering Design Can Be Expected
玄源明   先驱设计未来可期

123 THOMAS DARIEL   Design Is Signature
唐启龙   设计是你的签名

153 DANNY FANG   Create Love, Not More
丹尼·方   创造爱·宁爱勿滥

175 JURGEN BEY   Designed to tell the story of language exploration
尤尔根·贝   设计是诉说探险故事的语言

227 MICKEY HUANG   Paper Down, Mind Up
黄明强   放下图纸 思绪游弋

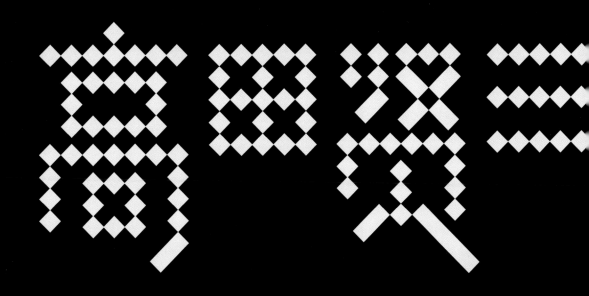

KENZO TAKADA   Dream Warms All

梦想给每个人温暖

KENZO TAKADA

高田贤三

1939年2月27日,
出生于日本兵库县姬路市。
著名时尚品牌KENZO(包括香水、化妆品及时装)的创始人。

高田贤三这位法国籍的日本时装设计师用自己的名字贤三命名的KENZO品牌,已经不仅仅是时装业的精品,在化妆品、香水领域也是大名鼎鼎。KENZO的时装不是那种标新立异的拔高,它有一点点传统,有许多热情的颜色,有活生生的图案,还有几分狂野。几乎每一款都能找到实际穿着的场合,但吸引力之强却使你绝不厌倦。

高田贤三年轻时就读于日本唯一的一家时装学校,他父亲对此曾竭力反对,因为当时在日本,还没有男人以缝纫制衣为业的先例,高田贤三本人也是时装学校中仅有的男性学生。1964年日本政府筹办东京奥运会,当时他居住的位于市中心的那幢大楼被划入用来兴建体育场地征地范围中。于是靠一笔35万日元的拆迁费,他买了一张去马赛的货船散客票。临行前,他把自己设计缝制的时装塞入五只大行李箱中,指望着它们会在异国得到赏识。在漫长的旅途中,远洋轮一路停靠许多城市。异域的风土人情在高田贤三的心中留下了无法磨灭的印象。他渐渐地领悟到以往的设计幼稚之极,索性把身边的五大箱旧作统统倒入了大海。

在巴黎,有一天,他惴惴不安地把几张草图交给设计师路易•费罗(Louis Feraud)看了,后者建议他不妨给《ELLE》投稿。当时《ELLE》已开始重视发掘新人的工作,并根据来稿中较好的设计草图制成成衣以Bon Magique品牌推出。高田贤三正是抓住了这个契机,打开了通向时装设计圣殿的大门。

他设计的首要原则是"自然流畅、活动自如",这是指结构,追求对于身体的尊重。高田贤三是第一位采用传统和服式的直身剪裁技巧,不需打折,不用硬身质料,却又能保持衣服挺直外型的时装设计师。他说:"通过我的衣服,我在表达一种自由的精神,而这种精神,用衣服来说就是简单、愉快和轻巧。"

他的灵感从印地安人、蒙古公主、中国传统图案与字样、土耳其宫女到西班牙骑士,就像跟他经历一段令人难忘的旅程,里面承载着各国绚烂夺目的民族之光,等待人们的发掘,尤其是他对于东方瑰丽而神秘色彩的偏爱,使得他能够将不同的民族特色融合在一起。

他把四季都想象成夏天,在颜色上变换着扣人心弦的戏法。具有民族特色的深葡萄酒红、艳紫、暗茄紫色、卡基和油蓝,是他经常使用的颜色,这些构成了流泻着温暖感觉的五彩缤纷的组合,而且又结合成更强烈的效果,燃放开来。

他喜欢猫、鸟、蝴蝶、鱼等美丽的小生物,图案往往取自大自然,尤其倾心于花。大自然的花朵、中国的唐装与日本和服的传统花样等,他使用上千种染色及组合方式,包括祖传的手制印花、蜡染等方法来表达花,从而使他的面料总是呈现新鲜快乐的面貌。

他的精神是遨游在东方和西方之间的,他的原创精神,在流行服饰中,代表了一个美丽的典范,在色彩多元化方面呈现快乐又简单的和谐。

高田贤三说:我用自己的方式传达和平与和谐的讯息,我希望人类的世界可以少一点自私,少一点个人主义;人与人之间多些宽容、互相分享和尊重。1999年,为表彰这位杰出的法籍日本设计师为人类和平所做的贡献,经过联合国33位大使的投票,高田贤三荣获了"TIME FOR PEACE"奖。这一奖项是由美国纽约一家非牟利性组织设立的,专为表彰那些为人类和平、理解与和谐贡献力量的艺人,并得到了世界的公认。高田贤三成为获此殊荣的第一位时尚大师。

这位带着一脸灿烂微笑、留着浓密的娃娃式长发、谦逊而幽默的艺术家在通往巴黎的成功途中也历经了暗淡而艰难的日子,但他的作品却始终没有丝毫的忧伤,就像雷诺阿的画一样,只有快乐的色彩和浪漫的想象。他因而被称作"时装界的雷诺阿"。

I HAVE A DREAM WITH ART
我有一个艺术的梦想

从很年轻的时候开始,我就拥有一个艺术的梦想。

在1960年之前,我从未想过出国,从1963年开始,我出国了,开心地畅游欧洲。我很奢侈地到了巴黎,在航行的游船里随意走动,我到了马赛,后来到了香港,那时的香港什么都没有,人们穿着很老式的中国衣服。香港之后去越南、新加坡、孟买、埃及、菲律宾等国家和地区旅游,之后我回到了东京。看到这么多不同的地方,看到不同的人不同的文化,是我非常难忘的,记忆也由此领略了世界各地不同文化的精彩。

刚到巴黎的时候,大约半年我的钱就用完了。在巴黎工作时,常思考作为一名设计师,我的身份识别是什么呢?之后发现我必须提供东方的内容,那就是日本的文化。服装设计蕴藏着巨大的发展空间,我很年轻,又很大胆,我愿意尝试着创造以前从未见过的。

1970年,服装的潮流推崇年轻崇尚自由,1978年,是服装设计的嬉皮时代,寻找着新自由。70年代的时装没有现在这么丰富的款式,那时一切都简单,消费者会很快地作出选择,而现在的时装多种多样,对于什么是前卫的什么是时尚的,消费者需要花费一段时间衡量才能作出选择,还往往是很难选择。80年代,我的服装设计能够成功,我要感谢媒体、感谢杂志。这时,我设计的时装,它不仅是服装秀了,而是必须是能够穿在身上的时装,扩大了消费者接受的范围。

1986-1991年期间,我希望有空闲时间,我不想工作了,这是转型期。于是我在巴黎住了很久,但是依然很喜欢日本的建筑、日本的庭院,我想把日本文化的禅意、诗意都带到巴黎。2000年我退休了,离开了KENZO。2003年我学习了画画,周游列国,我非常开心这样的生活形态。就在这期间,我认识了一个法国记者,他告诉我一个词,是:梦想。

是的,梦想,梦想就是坚持。

在服装设计方面,创意是最重要的。时装设计需要和合作伙伴成为一个很好的团队,时装的风潮容易波动起伏,但是不管商业最终如何发展,设计师需要始终如一地坚持创意设计,并且不间断地把这份工作从事下去。遇到困境,设计师必须接受挑战,事实上有时你会享受梦想实现的快乐。

Design by Kenzo Takada

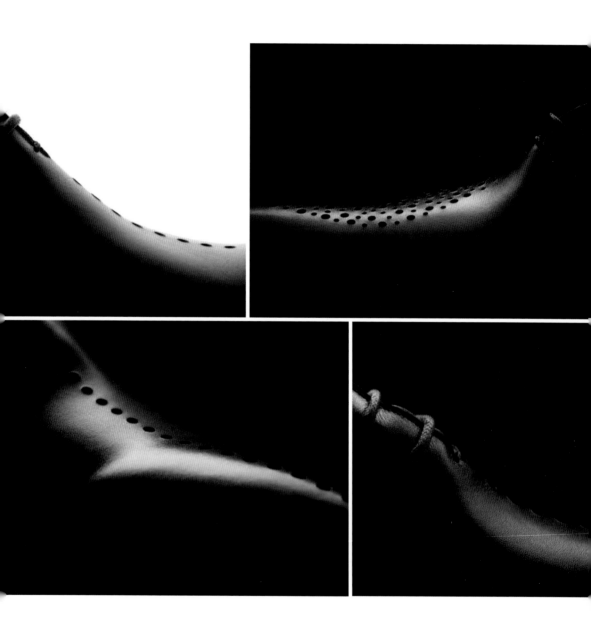

Design by
Kenzo Takada

INTERVIEW
KENZO TAKADA

高田贤三访谈

5只猫：有人说您像一块"艺术的海绵"，汲取东西方各种不同的文化素材，您的乐观精神和东方诗意美感也一直延续在您的时装设计中。其实也有很多东方的设计师带着他们的设计理念，希望被西方人所认可，而只有您成为最成功的亚洲时装设计师，您是如何使东方文化被西方接受的呢？

高田贤三：我觉得还是与时俱进的创作表现出了相当的展现力。从身份来讲，我属于日本也属于亚洲。我觉得能够被大众所接受的最主要的原因之一就是我把我自己和西方的文化融合在了一起。

5只猫：您被挑剔的巴黎人誉为"时装界的雷诺阿"，以自己名字命名的品牌也响誉世界，您是如何在艺术创作和商业运作上都做得如此优秀的？艺术家通常对商业不是很精通？

高田贤三：我在70年代完全忙碌于创作了，而到了80年代，比起创作我也许更多的是忙于广告宣传。80年代一起共事的管理者们将品牌进一步完善与推广。作为设计师的我还是以设计创造为主，关于营销方面，如果没有与专业人士结交，我想也不会取得成功。

5只猫：您的时装作品充分利用东方民族服装的平面构成和直线裁剪的组合，同时使用高纯度颜色面料和多色彩自由配组，形成宽松、自由的着装风格，这种设计风格与您个人的经历有关系么？

高田贤三：大致来说，是以日本和服以及亚洲各民族的服饰为基础。70年代的宽松样式，来自于和服的灵感，还有歌舞伎服装的颜色感觉，加上亚洲各民族的服装特点，通过自己的创造力在其中获得新的创作力量。这也就是颜色与样式等各种因素的大融合。

5只猫：KENZO创意总监玛哈斯称他和您都有游牧文化意识的存在，对异国情调的魅力充满好奇，所以才创造出KENZO许多具有异域风情的作品，您怎么看？

高田贤三：我觉得他很好地将他内在的那种异国情调融入了KENZO品牌当中。

5只猫：您的创意常呈现出一种混搭的画面，一件作品的面料和图案往往有不同的来源，有人说这其实是由于以前您经济拮据，不得不采用跳蚤市场上淘来的布料，久而久之就形成了这种独特的风格，这是真的么？

高田贤三：在70年代作品发表的时候，使用日本旧衣服（和服，浴衣）和其他在二手市场购入的材料，经过加工而后发表的事的确有过。花了一个月的时间乘船去法国所游历的这一切（接触到了各个国家的文化和传统）成为了激发我创造力的一个重要因素。

5只猫：您给人印象最深的是1978年的秀，您的秀场被装扮成一个马戏团,女骑手穿着透明效果的骑马装,您则骑着一头大象,非常幽默,让人看了非常开心,同您的作品一样,仿佛永远是带给人快乐的色彩和浪漫的想象,您是如何保持这种快乐设计的?

高田贤三:在我的字典中,有个词叫做"梦想"。我认为一个拥有梦想的祥和世界能够给每个人带来温暖。

5只猫:以您为代表,山本耀司、川久保玲等出生于日本成名于巴黎的时装设计师都为欧美本土文化吹入了一股清新而绵长的东方之风,也给东方本土"奋战"的时装业同行以莫大的鼓舞与信心,您觉得一个设计师要获得成功,需要拥有哪些条件?

高田贤三:我觉得巴黎的时尚作品非常追求季节的特色。在新作发表当中出现的作品,一定要是平常可以穿着的服饰。创作与商业的确有相悖之处,如何将这对矛盾调和是很重要的。

5只猫:您是在日本东京文化服装大学学习 (Bunka Fashion College) 毕业后不久就去了巴黎,那时候您才25岁,您觉得大学教育对您的设计起到了什么样的作用?巴黎的学习又带给了您什么?

高田贤三:我从小池老师那里学到了很多。刚才我也提到过,在去巴黎的时候,"请坐船去"就是小池老师对我说的。在巴黎我并没有去上学,在我的新作发表时,我觉得那次坐船旅行当中,我的所见所闻对我的创作有相当大的影响。

5只猫:我们得知当初您要做设计时,家庭是反对的,这是事实么?是什么机缘使您产生了对设计的兴趣呢?

高田贤三:当初受到的反对,并不是针对于"从事创造"。关于去东京学习,的确多少遇到过些阻力,但后来还是得到了大家支持。

5只猫:您怎么看待中国的设计?中国对主流奢华品牌的关注度越来越高,却缺少自己优秀的品牌,您有什么建议带给中国的设计师么?

高田贤三:我觉得中国的设计师从今往后会越来越向海外方面发展。欧洲的很多设计师希望承揽中国的设计项目。把握好创造与经营之间的平衡,开创自己的事业并成功地将其运行,这是我的期望。

5只猫:您的KENZO品牌涵盖了女装、男装、女士香水、男士香水、护肤、家居等方面,而您也在巴黎举办过画展,可以说您的艺术才能不仅仅局限在服装设计上,是什么让您成为一个多方位的艺术家呢?

高田贤三:现如今,我本人跟KENZO品牌没有什么关系。画展是我新萌生的一个想法,通过这个方式能够跟各行业形成合作,进而实现继承和发扬世界各国的传统文化,可以说这是我以后人生计划中的一环吧。

5只猫:作为一名享誉时尚界的设计师,您怎么看待时尚?

高田贤三:我觉得时尚界是我的梦之所在。特别是如果能够通过这种创造,向外界传达出各种信息的话,那将是很有价值的事情吧。

5只猫:您的时装里有一种让人无法拒绝的嬉皮色彩,而这种色彩又始终交织在浓郁的东方风情里,令人感受到一种说不出的喜悦和轻松,从1970年成名到现在,您是如何保持这种风格的?让自己的作品时刻拥有竞争力的秘诀是什么?

Kenzo Takada 高田贤三:人类有五感。我认为,要将这五种感觉发挥至极,表现出个人的感性一面这才是对于一个作品最为重要的。

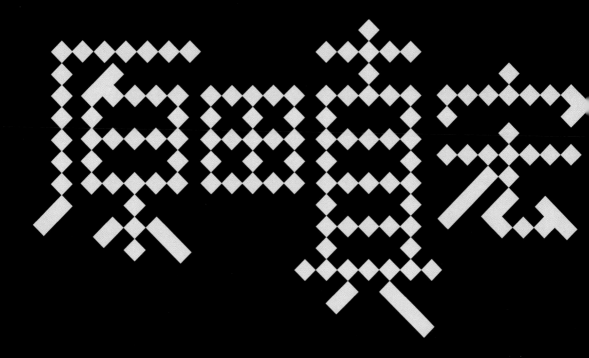

MASAHIRO HARADA
The Age of Architecture=Art Is Coming

建筑=艺术的时代即将到来

MOUNT FUJI ARCHITECTS STUDIO

富士山建筑师工作室

Masahiro Harada + MAO

原田真宏 + 原田麻鱼

Awards 奖项	
2007	The Barbara Cappochin Prize for Architecture Best international works
2007	获芭芭拉·卡布其建筑奖最佳国际作品奖
2008	AR Awards for Emerging Architecture 2008" RAINY / SUNNY" Honourable Mentions
2008	作品"雨天/晴天"获2008AR新兴建筑佳作奖
2009	AR Awards for Emerging Architecture 2009 (UK) "PLUS" Commendation
2009	作品"加"获英国2009AR新兴建筑表彰奖
2009	Design Vanguard 2009 ARCHITECTURAL RECORD (U.S.A) TOP 10 architect in the world
2009	获美国2009设计先锋"建筑实录"世界十大建筑师称号
2010	RECORD HOUSES 2010 Architectural records (U.S.A) "Tree house"
2010	作品"树屋"获美国"记录的房子"2010建筑实录奖
2010	LEAF AWARDS 2010 Residential Building of the Year -Single Occupancy- (UK) "PLUS"
2010	作品"加"获英国"叶奖"2010年度单人住房奖
2010	Chubu Architecture Award 2010(Japan)"PLUS"
2010	作品"加"获日本中部建筑奖

"DIALOGUE" BETWEEN OLD AND NEW "SUBSTANCE"

新旧"物质"间的"对话"

We do not subscribe to the assertion that "the city is a problem and architecture is the answer." That point of view is a pure product of modern architectural theory, which as such weighs very heavily on today's architectural education programmes: What are the problems running through the city? What answers can architecture offer them? School trains us in the acquisition of this method of questioning. Student evaluation is based on this conceptual and rational system of question and answer. And it is doubtlessly relevant, if limited to academic training; architecture on paper, devoid of substance, remains at a level of abstract purity that allows it to theoretically resolve the problem posed by the city. But with real architecture it is quite anther matter. Indeed, even when it is designed as a pure answer, architecture realised, from the moment it imposes "mass" and becomes a built object, never manages to get beyond the "city=problem" equation. Because many architects have not grasped the obviousness of this, an incalculable number of buildings have sprouted in the urban landscape through the conscious application of the lesson learned: "problem-solution." Unfortunately, the legitimate and equitable "answer" expected often winds up being nothing more than deplorable "urban filler". For in using this approach, the concrete situation of the city is rendered abstract, theorised and formalised as problem and turned into a set of logical systems which will in turn administer a logical architectural answer. It is useless and unsightly to reintroduce these relationships defined through the filter of conceptual labels into the material world in the form of buildings. the resulting built architecture is merely a superfluous residue. We are doubtless the first generation to become aware of the reality of modernism's limits. We sincerely and conscientiously avoid dealing with architecture through concepts as much as possible. For us, the city is from the outset imbued with "substance," and the architectural process is the creation of "substance". Therefore, we seek to manipulate these concrete relationships, as they are, in all their concreteness. The relationship between pre-existing city and future architecture is never envisaged in a unilateral way, as one would do when bringing an answer to a question, but rather as a continuous and balanced "dialogue" between the old and the new "substance." This is what makes our point of view so childlike. To act upon things simply, so they will actually become what one would wish for.

我们并不赞成"城市是问题，建筑是答案"这种说法。这种观点是现代建筑理论的纯粹产物，而它在今天的建筑教育大纲中占据着举足轻重的位置：城市的问题何在？建筑能够给出何等答案？学校训练我们用提问法来寻找答案。对学生的评价也建筑在问答的概念和理性系统的基础之上。如果仅限于学术教育，这毋庸置疑是有所关联的：纸上建筑，无物质实质，停留在抽象纯粹的层面，允许我们从理论上解决城市抛给我们的问题。但是，真实建筑却是另一回事。实际上，即使它被设计成一个纯粹的答案，而建筑使答案变为现实，但是从它施加于"大众"成为建筑实体时起，便永远无法逾越"城市＝问题"这个等式。因为很多建筑师没有抓住其中显而易见的东西，无数的建筑在城市景观中拔地而起，都是对"问题-解决"模式的"学以致用"。不幸的是，人们所期待的合理的对等"答案"往往最终不过是可悲的"城市填充物"。利用这种方法，城市的具体状态被认为是抽象的、理论化和形式化的，问题变为一个逻辑系列，反过来管理逻辑性的建筑答案。将通过概念标签过滤形成的上述关系重新以建筑的形式引入物质世界是徒劳和丑陋的，最终产生的建筑只不过是多余的残渣。我们无疑是对现代化在现实中所受的种种限制产生感知的第一代人。对于我们来说，城市从最开始就被灌输了"物质"，而建筑的过程就是创造"物质"的过程。因此，我们寻求对上述具体关系的操纵，因为它们是完全具体的。已经存在的城市和未来建筑之间的关系永远不能是单边的，就像有人在回答一个问题一样；而应该是新旧"物质"之间连续而平衡的"对话"。这就是使我们的观点如此天真的所在。对事物做出简单的反应，它们就会真的变成你所期望的样子。

INTERVIEW: MASAHIRO HARADA
原田真宏访谈

5只猫：我们都知道日本著名建筑师安藤忠雄，他从未受过正规科班教育，却也开创了一套独特、崭新的建筑风格，而我们知道您是芝浦工业大学工学部的客座教授，是学院出身的设计师，您怎么看待艺术和教育之间的关系？

原田真宏：技术方面另当别论，和其他艺术一样，属于艺术的建筑领域不能教条式的教，重复使用规定的风格有什么意义呢！建筑是通过感知世界并抽象化产生的。换句话说，我们通过每天认识世界，每天创造出一个全新的合理的风格。这样每天面对世界并进行抽象化的建筑师背后的姿态才有实质上的教育效果。

5只猫：日本设计师原研哉说过，"设计与其说是在制造物品，不如说是发现生活的智慧。"那么如何将生活的智慧展现给大众呢？

原田真宏：设计是新的合理性的发现。如原先生所言，设计不是新奇而象征身份的商品。我们的建筑也是一样的。比起杂技的新奇，我们致力于把未发现的合理性实现于建筑上。传统建筑是按传统的"生活方式"建造的，反过来讲传统建筑制约着"生活方式"按旧习惯延续。可以说是建筑强迫习惯从属。这里没有自由。我们要做的是，诱导居住者让其自主发现适合自己的"生活方式"。所以说，比起习惯的"住宅"，寻找适合生活的"地形"更接近我们的目标。

5只猫：您的作品《树屋》利用"向心趋势"将一间原本缺乏自然光与私密性的公寓改造成空间合理倍觉舒适的住宅，这种本身自然条件不好的住房改造有很强的现实意义，请谈谈您设计之初的构想。

原田真宏：确实周围被邻居包围的树屋，不能说有很好的自然环境。然而，从邻居家的高低不平，丘陵地形来看，屋顶空间还是有享受好环境的可能性。我们选择的建筑树木与自然树木一样，决定的几何学是最适合环境的。还有，采用了接近单房的方案，将有限的自然要素贯穿于室内全部，为了做到求新性强的结构，采用了在中心位置放置大黑柱的"田字形方案"的日本传统建筑模式，不但理性现代，而且给予安全感。

5只猫：现在中国有很多城市、很多开发商在建一些外观与众不同的建筑，它们通常被冠以"前卫建筑"的称号。比如北京的央视大楼，广州的大剧院，电视塔"小蛮腰"，您怎么看待这种潮流？

原田真宏："发现适合新世界的新的合理性建筑模式"是好的事情，这有点儿像20世纪80年代的日本。中国由于经济发展目前犹如世界前卫建筑试验场一样。应注意的是，没有新的合理性只是"新奇"的话像吃惊箱一样很容易被厌倦。而且，建筑是位于城市景观中心，不能像吃惊箱一样随便处理掉。所以，居住在那个城市的每个人都要学会区分"新合理性"和"单纯的新奇"的区别。

5只猫：时下人们对居住环境的环保和健康要求也越来越高，作为设计师应该如何去设计更生态更低碳的建筑，而不仅仅将保护环境停留在口号上？

原田真宏：作为一名建筑师重要的作用是寻找合理的解决方案。到目前为止建筑追求的是服务于人类有效果的象征，但从现在开始不光是人类，更应该重视遵从大自然的呼声。以大自然的呼声作为最重要背景的建筑模式目前还没有发现。我的建筑活动就是为了要发现它，实现它。

5只猫：作为一衣带水的邻邦，中国和日本在文化上有着源远流长的关系，如何保护和发展自己独特的文化成为中日都在思考的问题，作为一个设计师，您怎么看待西方设计文化对东方的影响？

原田真宏：20世纪西方的"人类文明的对自然的存在"已

经进行了有限的展示。亚洲的文化，我们有"对自然亲和的文明的存在"，我认为这将成为日益严重的环境问题，是这个世纪以后文明的重要思想。

我们好不容易拥有的文明美德，就这么简单的被西方以人类为主文明背景的过去资本主义所吞噬是非常惋惜的事情。将现实抽象化理解的理性思维是以人为本的西方文明资产之一，对于想把非合理的世界从黑暗中解救出来的世界人类来说是一个重大的成就。但对于如何持久的保护全球环境的问题，坦率地说有些不足。每个建筑师应有能力与抽象化之前的"原始自然"相处互融。

谁能活跃于这个部分，不光是日本及亚洲，乃至全世界所期待的事情。

5只猫： 现在的很多设计师在作品中融合了中西元素，融合中西与保留个性之间应如何平衡？在某种设计流行趋势下，又怎样让自己在创作中保持自己独特的想法？

原田真宏： 国际化设计潮流有价值的原因只是因为它保持了本土个性。正如现在不管去世界哪个地方城市景观都很类似，国际化设计好像席卷了整个世界，现在设计背景已经改变了，现在有价值的是有地方特色的设计。

5只猫： 在快节奏和繁忙的都市生活中，设计师该如何保持敏锐的感受力和丰富的创造力？

原田真宏： 对我来说，是海泳，爬山（笑）。
说真的，吸收行业外空气，注意不要自我文化中毒也是很重要的。

5只猫： 时下有越来越多的设计师参与到"跨界"设计，您怎样看待设计和其他艺术之间的界限？

原田真宏： 历来，建筑被定位为各种艺术的综合体。比起只是实现功能的廉价建筑，现代建筑更为突出这一点，这样一来，艺术实现了融入建筑的可能性。
因为环境问题，建筑将从"量"转化为"质"的价值，可以再次实现建筑与艺术的结合。设计技法在建筑与艺术中间有很多共有的部分，所以我想新的建筑＝艺术的时代即将到来。

5只猫： 还记得您的第一件设计作品是什么？设计给您的生活带来了什么？

原田真宏： Tree House 是我们的出道作品。当时的预算是一台大众车，也就是仅用 1500000 日元工程费实现的。然而，这个小建筑里蕴含着我们所有人的未来的设计思路。这个建筑开辟了我作为建筑家的道路（对于建筑家来说，出道作品可以说是决定今后设计人生的很重要的东西）。我想设计给我带来了人生的精神自由。

PLUS

+

Function : Villa+Petit hotel
Location : Shizuoka, Japan, 2009
Site Area : 988.58m²
Building Area : 232.77m²
Total floor Area : 380.44m²
Number of stories : 2 story+1basement
Structural system : Reinforced Concrete

功能：别墅+客栈
位置：日本静冈，2009年
占地面积：988.58m²
建筑面积：232.77m²
房屋总面积：380.44m²
楼层数：2层+1层地下室
结构：钢筋混凝土

The site locates on mountainside of Izu-san, where Pacific Ocean can be looked down on the south. The untouched wilderness, covered with deciduous broad-leaved trees such as cherry trees and Japanese oaks, gives little level ground. But we saw faint glimmer of architectural possibility along the ridge. The architecture would be used as villa for weekends.

建筑地点位于伊豆山山腹，南邻太平洋。这里是人迹罕至的世外桃源，崎岖的地表上生长着樱桃树、日本橡树等落叶型阔叶林。但我们在山腹地带发现了一线曙光，这里可以用来建造周末别墅。

I didn't want to just form the undulating landscape dotted with great trees as normal, nor design an elaborate architecture bowing down to the complex topography. What sprang to my mind is a blueprint for an architecture which is perfectly autonomous itself, at the same time seems to emerge as an underlying shape that the natural environment has been hiding. It's abstraction of nature, to say.

我不想仅仅是将点缀着参天大树的崎岖地表变为普通的建筑地面，或者屈服于复杂的地表形态，让设计变得精巧。我脑海中的建筑蓝图应当是完全独立自主，同时看起来又揭示了被大自然隐藏的隐秘地形。可以说是抽象的自然。

The architecture was realized by crossing two rectangular parallelepipeds at very right angles. The lower one contains private rooms and bathroom, and sticks half of the body out to existing narrow level ground. The upper one incorporates salon and kitchen, and lies astride the lower one and the mountain ridge. It almost seems like an off-centered cross pinned carefully on natural terrain.

低洼区域包括私人房间和浴室，房屋的半身延伸到已有的狭窄平地。高隆地带骑在低洼地带和山脊上，由沙龙和厨房组成。看起来就像是别在天然地表上的一个偏离中心的十字。

One axis of the cross stretches toward the Pacific Ocean on south, and the other, the forest of Japanese oak and some white birch on west. The rooms in the lower structure and terrace on it enjoy broad vista of the sea and blue sky. And gentle shade of natural forest embraces the space in the upper one. Water-polished white marble (cami #120) was chosen as interior finishing material. It glows softly like Greece sculptures to blend blue light from the south and green light from the west gradationally, thus creates delicate continuous landscape of light which suggests the character and usage of the space. Exterior is also finished with white marble. The surface get smoother as it approaches to the southern/western end till it takes mirror gloss (cami #1000) at the ends. The southern end of white cross melts into the blue of sky and sea, and the eastern end to the green of forest.

十字的一个轴向南延伸到太平洋，另一个轴延伸到日本橡树林和西边的一小片白桦林。站在低层结构的房间和房间上的平台上，可以远眺大海和蓝天。隐隐约约的树林环抱着上层结构。内饰材料采用的是水抛光的白色大理石（cami 120号）。这种材料散发出柔和的微光，就像古希腊的雕塑，渐进地融合了南方的蓝色光线和西方的绿色光线，从而创造出一种连续而柔美的线型，指明了空间的性质和用途。外部饰以白色大理石。越接近南边或西边，表面越光洁，末端采用的是镜面抛光（cami 1000号）材料。白色十字的南端融入蓝天碧海，东端隐入绿色森林。

Abstraction is nothing to conflict with nature here.
Carved out of nature, it never stops being a part of nature itself, however highly abstracted. Never relativizes the nature with its foreignness, nor generate contradiction to settle for being "artificial nature" by giving up being abstract and mimicking the nature.
The abstraction inspired by Mother Nature defines the nature itself, and still, stays natural.
That's what I wanted from this abstraction and architecture.

抽象和自然在此处合为一体。
从自然中雕刻，绝不会脱离于自然本身，但却具有很高的抽象性。不要凸显自然的不同，也不要为了创造矛盾而妥协于非抽象、拟自然的"人造自然"。
大自然母亲给我抽象的灵感，又以此对自然进行定义，同时它又回归自然。
这就是我想从抽象和建筑中得到的。

(Harada Masahiro / Mount Fuji Architects Studio)
(原田真宏 / 富士山建筑事务所)

TREE HOUSE

树屋

Function : Private residence
Location : Tokyo,Japan,2009
Site Area : 162.69m²
Building Area : 78.76m²
Total floor Area : 80.45m²
Number of stories : 2story
Structural system : Wood

功能：私人住宅
位置：日本东京·2009年
占地面积：162.69m²
建筑面积：78.76m²
房屋总面积：80.45m²
楼层数：2层
结构：木质结构

This is a house for young editor's couple in typical residential area of north part of Tokyo.
这栋住宅是为东京北部典型住宅区的一对年轻的编辑夫妇设计的。

The site is surrounded by neighbor's houses closely for all directions. It seemed difficult to open the house to horizontally. Normally this site-type is hated for housing for the reason of problem of shortage of natural light and privacy. But we found one unique potential with this dusky site. That is "centripetal tendency" by limitation to horizontally.
四周房屋鳞次栉比，横向拓展十分困难。一般而言，这种类型的地块由于采光和私密性较差而不受设计师欢迎。但我们发现了这块"灰色地带"的特别之处。那就是在横向受限的同时可以向上发展。

So we selected the Polar-coordinate-system as geometry of architecture instead for Cartesian-coordinate-system which is generally used in architectural design. The rule is very simple. Each LVL-frame was rotated and reproduced by 11.25 degrees. And every frame is higher 55mm than next one. As a result, the rational strong structure which looks like big tree has been realized.

所以我们在建筑几何学上放弃了建筑设计中普遍采用的直角坐标系，而采用极坐标系。规则十分简单 我们将每个层积材框架旋转，形成11.25度的角度。每个框材都比下一个高出55毫米。从而用合理又结实的结构搭建出了大树的造型。

The main space is divided to 4 different characterized spaces by this big Tree-like column. All spaces have different composition by height and width and different light amount. So we imaged actions along these characters of spaces. For example, high and well lighted space is for dinning, low and dim cozy space is for sleeping.. In this house people will find his favorite space by himself, not by order from named rooms with functions.

核心区域被树形柱分成四个各具特色的部分，这让我们可以根据各部分的特点分配区域内的活动。例如，可以在高亮区域用餐，可以在低暗温馨的区域睡觉。人们可以在房子内找到自己喜欢的区域，而不仅局限于房间的功能和名称。

But beyond my many words, this old song express my concept simply and correctly.

除此之外，我想用这首老歌简单又贴切地表达我的设计理念：

"Under the spreading chestnut tree.
There we sit both you and me.
Oh how happy we will be.
Under the spreading chestnut tree."
(UNDER THE SPREADING CHESTNUT TREE)

"在枝繁叶茂的栗子树下

坐着我和你

哦，我们是如此开心

在枝繁叶茂的栗子树下 "

（《在枝繁叶茂的栗子树下》）

I wonder that we could realize the space sang by this familiar song in Tokyo. (Masahiro Harada)

我想我们可以将东京这首耳熟能详的歌变为现实。

（原田真宏）

Now I remind that the way to divide the space by symbolic center column is very traditional way in Japan. We can find many old houses by same system. Maybe this is one reason why old grandmothers of neighbor's like this very modern house.
In the way, this geometry succeeded to change the roof to interesting terrain. Normally it is difficult to get comfortable garden in this kind of situation. But there exist sunny and opened terrain on the roof.

现在我想起，用象征性的中心柱体分割空间的方法是日本的一种传统方式。我们可以看到很多老房子都是这么建的，也许这也是邻居的老祖母喜欢这所极具现代感的房子的原因之一。这种几何构造通过这种方式将房顶改造成极具趣味的空间。一般来说，在这种条件下很难建一个舒适的花园，但房顶提供了阳光充足的开阔区域。

NEAR HOUSE
近水楼台

Function : private residence
Site Area : 66.42m²
Building Area : 37.65m²
Total floor Area : 75.30m²
Number of stories : 2story
Structure : wood frame
Design Date : 2008.9-2009.4
Construction : 2009.7-2010.2
功能：私人住宅
占地面积：66.42m²
建筑面积：37.65m²
房屋总面积：75.30m²
楼层数：2层
结构：木框结构
设计时间：2008.9-2009.4
工期：2009.7-2010.2

Nearness -the other side of smallness-
近,小的另一面。

The site is located in a dense residential area in Tokyo where inhabitants are allotted only limited lands. A landowner sold the already limited land to two owners by dividing it. Quite small. Also, irregular in shape. Viewed from above, it looks just like vertically-long flag and pole. Perimeter is long for its size, and the walls and the trees of the houses next-door stands very near. But they don't give oppressive feeling. Rather, the closeness –the houses look just like they're cuddling each other- creates some intimacy.
该地块位于东京的住宅密集区,居住者分摊面积有限。地块所有人将本就有限的土地分割后卖给两位买家。一隅之地,奇形怪状。从上方俯视就像长长的竖着的旗子和旗杆。周长相对于大小来说很长,临近房屋的墙壁和树木近在咫尺,但并没有压迫感。反而距离的接近——房子互相依偎着——创造出一种亲密感。

The site is small, all right. But when you change your point of view, its "nearness" can be constructive advantages. We wanted to bring the best out of it.
地块面积小,其实还好。若换个角度想,这种"近距离"能够成为建筑的一种优势。我们希望尽其所用。

The professions of the owner couple and issue of building coverage divided the architecture into two small buildings and a court (*1). The buildings sit near each other. So the court seems more like a part of building that connects two room spaces.
考虑到屋主的职业和房屋面积的问题,我们将其分成两栋小的房子和一个院子(*1)。两栋房子非常接近,而院子就像是连接其中的桥梁。

Each building has bonded wood (larch) post-and-beam placed at 450mm intervals. The narrowness of the interval defines whole scale of architecture. Near House is more a furniture or a product than architecture (*2). It interacts with people so closely. And that's why we have the surfaces of bonded wood, softwood MDF and mortar finely textured like that of peach. Such delicate finish can only be possible thanks to the size, which would normally be considered as disadvantage. The house doesn't have so much space. But the high-resolution world beyond the compass of architecture creates a sense of depth in different spheres.
每栋房子的梁柱结构都采用胶合板(落叶松)材质,间隔为450mm。狭小的间隔为建筑的局促下了定义。近水楼台比起建筑来更像一件家具或一个产品(*2),它可以与人们近距离互动。这也是为什么我们采用像桃树纹理一样的胶合板、软木中纤板和灰浆做表面,这种精巧的装饰只能归功于通常被认为是缺点的狭小的空间。房子不需要有多大,建筑之外的高分辨世界为不同的领域创造了深邃感。

By finding "nearness" in "smallness" and making the most of it, the house transcends the realm of architecture, and takes a step into the world of product. In that sense, Near House can be considered as "Something near to be a house".
(Masahiro Harada)

通过在"局促"中找到"亲近"并最大化利用这种亲近,房子超出了建筑的范围,成为产品的世界。从这种意义上来说,近水楼台可以被称为"近房屋的东西"。
(原田真宏)

(*1: The areas that allow construction are the "flag" part, and the "grip" part of the "pole" that is originally designed as parking and has some more width than the "pole" itself.)
(*1: 可以施工的区域是"旗子"的部分,而"旗杆"的"手柄"部分最初被设计为停车地点,比"旗杆"本身的宽度要宽一些。)

(*2: Typical Japanese mini storage shelf is 450mm in width and 15mm in board thickness. Near House is almost like consisted of so many of this.)
(*2: 标准的日本迷你储物格板为450mm宽,15mm厚。近水楼台几乎满是这些储物格。)

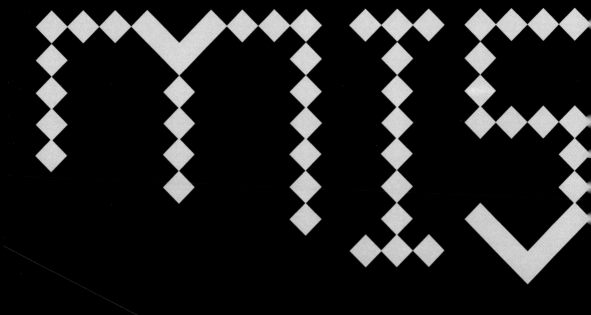

MI5 All about Nature

任何的创作都是回归自然

INTRODUCTION
MI5

MI5 is a team of Manuel Collado Arpia, Nacho Martín Asunción since 1999 which focuses on researching unexplored projectual strategies, our work has been recognized, awarded and published on several occasions.
MI5 作为曼努埃尔•哈比和纳乔•亚松森领导下的工作团队，自1999 年以来一直致力于研究尚未开发的项目策略，设计项目曾多次发表，并受到业界肯定和嘉奖。

We are invited to: VII Venice Architecture Biennale, Europe VI Burgos, Liege and MOPU, COAM Recent Work Foundation, FRESHMADRID COAM Foundation, Madrid Social Housing RIBA (London), 100% Crude COAM Foundation, Building Dwelling Thinking IVAM ...
MI5 曾受邀参加：第七届威尼斯建筑双年展、欧洲第六届布尔格斯展、Liege 和 MOPU、COAM（西班牙马德里的一所基金会）"近作"基金项目、COAM "新鲜马德里"基金项目、英国皇家建筑学会主办的马德里社会住房项目（伦敦）、COAM "100% 纯天然"基金项目、西班牙瓦伦西亚现代艺术博物馆主办的建筑居住思考展……

They have been teaching in Architectural Polytechnic Universities of Alicante, Alcala de Henares and Camilo Jose Cela, Fashion Design at the European Institute of Design (IED Madrid) and Visiting Member at the Architectural Association in London, in addition to having participated in several workshops, juries and lectures.
MI5 的教学足迹遍及阿利坎特、阿尔卡拉德埃纳雷斯和卡米洛何塞塞拉的建筑理工大学、欧洲设计学院的时尚设计专业，并在伦敦的建筑学会担任访问学者，除此之外，还广泛参与工作室、评委会和讲座等活动。

They have been awarded in 2010 with the "Annual Architectural Design Award" for Rivas Vaciamadrid Youth Center and "Annual Outstanding Designer Award" to Mi5 Arquitectos for the chinese magazine Modern Decoration.
2010 年，MI5 设计的里瓦斯马德里青年中心荣获"年度建筑设计大奖"，MI5 的建筑师更被中文杂志《现代装饰》评选为"年度杰出设计师"。

THINKING IN THIS TIMES
在这个时代思考

MI5´s profile was conceived since the turn of the century as a suggestive and innovative practice midway between construction and research.
MI5在世纪之交，被公认为是非常有建设性和创新性、兼顾建筑和研究的设计公司。

Our work focuses on the search of alternative and not very explored public management strategies, exploiting fantasy as a powerful, dynamic motor of social processes.
我们的工作核心是另辟蹊径，研究尚未开发的公共管理策略，发掘神奇，使其成为推动社会进步的活力和强大动力。

Interested in participating in the construction of the signs of our era, we conceive our projects as the erection of identities of collective recognition, capable of redefining and strengthening the geographical and human territory where they stand.
我们兴致勃勃地参与标志着我们这个时代的建筑项目，将我们的项目视为集体认知身份的建树，能够重新定义和强固其所在的地理和人文领地。

We work with the efficiency of our projects´ communication strategies as a proposal which will intensify their social viability. Our work method is supported by proactive policies, with very personalized relationship protocols for each user and his environment.
我们高效工作，奉行加强项目的社会生存能力的沟通策略。我们的工作法则得到积极的政策支撑，针对每个用户和具体的环境确立非常人性化的关系法则。

One of the missions of the projects, conferences, essays and teaching undertaken by the studio during the last years, has been rethinking concepts such as Public, Youth, Outskirts, Fashion, Technology…
过去的数年中，公司接手的项目、会议、文章和教学任务之一就是对诸如公共、青年、郊区、时尚和技术等概念进行重新思考。

INTERVIEW: MI5
MI5 访谈

5Cats: When people talk about Spain, great names such as the literary giant Cervantes, painter Picasso, tenor Placido Domingo always make the conversation. What Spanish designs move people are the romance and passion, like in your work Rivas Variamadrid Youth Center": striking designs with glamorous and bold colors and quirky shapes. How did you come up with the idea of that space at the beginning? Where did you start in your design? How did you start?

5只猫:提到西班牙,文学巨匠塞万提斯、画家毕加索、男高音歌唱家多明戈为世人耳熟能详,地杰人灵。西班牙的设计打动人心之处往往在于她既浪漫又富有激情,你们的设计作品《巴西亚马德里青年中心》就体现得非常明显:冷艳奔放的色彩和稀奇古怪的造型引人瞩目,在设计之初你们如何构思这个空间?设计从哪儿开始的?是怎么样开始的?

MI5: In the design process of Rivas youth center we did not have any influence of the Spanish classics. The passion in this case came from the youngsters who participated in the open design process. The power belongs to a group of young people who knew how to make signs of their needs and identities. Then our work was how to transform identity signs in a building.

MI5:设计巴西亚青年中心的时候我们没有受任何西班牙经典文化的影响。当时激情来自于参与公开设计活动的年轻人。作品的力量源于一群年轻人,他们知道如何将他们的需求和身份化为标志。然后我们的工作就是怎样将身份标志化为建筑。

5Cats: I believe a lot of designers like me are full of curiosity: What is the realization process of this space like? How does one design start from the drawing and finally become true phase after phase? In this process, what do you feel?
5只猫：相信很多的设计师和我一样充满着好奇心：这个空间实现的过程是怎么样的？一个设计，怎样从图纸通过层层环节最终实施完成？在这个过程中，你们的体会是怎样的？

MI5: The design was based on a participatory process, where the users at the beginning were given a rigid shoe box. Then with the youngster we decide to dynamite and break the geometric boundaries of this box creating a process of space customization.
MI5：这次设计是一个参与的过程，最初使用者拿到的是一个呆板的鞋盒子。然后我们和年轻人一起决定把盒子的几何边缘炸掉摧毁，创造一个用户化的空间流程。

5Cats: In your opinion, what is the most important in one design?
5只猫：你们认为，设计中最重要的是什么？

MI5: The most important thing is to get a design that clearly reflects the identity arisen between your client and your feelings.
MI5：最重要的是让设计清晰反映在你的客户和你的感觉之间生成的身份感。

5Cats: With the economic globalization, different cultures in different parts of the world gave birth to the clash of classicalism and modernism, and other art schools; while China stands for harmony between different cultures. So faced with different geographical and cultural clients, how to make them accept your designs?
5只猫：在经济全球化的今天，全世界不同地区不同文化，催生了古典主义和现代主义等艺术流派的碰撞；而中国主张不同文化之间和谐共存。那么面对不同地域和文化的客户，如何使他们接受你们的设计？

MI5: We always like to build bridges between the identity feelings for each country, we can't design without such linkages. Then we enjoy playing and creating with these feelings in many different ways, trying not to fall into clichés. We are always open to the wishes of each client to enhance them and turn them into constructed realities. All these things finally are very well accepted by a diverse audience.
MI5：我们总是喜欢在每个国家的身份感间架起桥梁，失去了这些联系，我们就无法设计。然后我们喜欢带着这些感情用多种不同方式玩乐和创造，尝试不落俗套。我们通常对每位客户的心愿都保持开放心态，将它们升华并变为真实的建筑。上述所有最终都会被不同的观众欣然接受。

5Cats: Your design is dynamic, full of elements of fashion, reflecting the mentality of young designers. As a young

designer, what role does traditional Spanish culture play in your design process? Did the conflict of traditional culture and modern culture ever confuse you?

5只猫：你们的作品活跃着动感、时尚的元素，反映出年轻设计师的心态，作为年轻的设计师，您们在设计过程中，西班牙传统文化究竟发挥了怎样的影响？传统文化与现代文化的冲突，对您们是否产生过困扰？

MI5: In our case there is not really a conflict between traditional and modern, because we like to use any kind of ingredient in our language. The problem is that in our recent works, the client belongs to a young and peripheral context, where there are fewer opportunities to use traditional elements.

MI5: 我们的案例中并不真正存在传统与现代的冲突，因为我们喜欢使用我们语言中的任何一种成分。问题在于我们最近的作品中，客户偏年轻和边缘化，因此很少有机会运用传统元素。

5Cats: The famous Spanish designer Antonio Gaudi said, "Creation is the return to nature". How do you understand creation's return to nature?

5只猫：西班牙著名设计大师安东尼奥·高迪说过"创作就是回归自然"，你们怎么理解创作的回归自然？

MI5: We are nature any creative process is returned to nature.

MI5: 我们是自然的，任何创作的过程都会回归自然。

5Cats: What qualities shall an excellent contemporary architecture have? What kind of design can be called a "global pioneering design"?

5只猫：优秀的当代建筑设计应具备怎样的特质？怎样的设计才称得上是"全球先锋设计"？

MI5: Excellence in contemporary architecture is an intelligent mix of creativity, research and technology.

MI5: 当代建筑的卓越之处在于它是对创作、研究和技术的智能融合。

5Cats: Low-carbon design is popular in the world. While some people said: low carbon is a lie. What is your opinion?

5只猫：低碳设计正风靡整个世界。可有人也说：低碳是一种谎言，你们持怎样的观点？

MI5: Sustainable design can never be a lie, what happens is that is no longer original talking about sustainability, because it must be understood as a normal process, such as digitization, or electrification.

MI5: 可持续设计永远不会是一个谎言，只是我们已经偏离了原来对于可持续的探讨，因为我们必须将其理解为一个正常的过程，如数字化或者电气化。

5Cats: When did you start to like art? Who are your favorite artists? What influence do they make on your art road?
5只猫：你们是什么时候喜欢上艺术的，最喜欢的艺术家有哪些，这些艺术家对你们的艺术道路有着怎么样的影响？

MI5: Architects are related to art from the beginning of our training, is essential to know the work of other artists. We are especially interested, considering main stream artists, in the work of Anish Kapoor, Olafur Eliasson, and Damian Hirst.
MI5：建筑师从我们开始接受培训起就与艺术有关，了解其他艺术家的作品十分重要。主流艺术家中，我们尤其对阿尼什•卡普尔、奥拉维尔•埃利亚松和达米安赫斯特的作品十分感兴趣。

5Cats: Are you are now living in Spain? Which countries and cities have you been to? Does international experience make a difference on your design? Which cities impressed you most?
5只猫：你们现在居住在西班牙么？到过哪些国家和城市？不同国家之间的游历对你们的设计历程有哪些影响？印象最深的城市是哪个？

MI5: So far we have managed to travel to all the continents. Like art, travel is in the base of our training.
MI5：到目前为止我们已经游遍了所有大洲。旅行就像艺术，也是我们的一门基础课。

5Cats: Please give your interpretation on the trend of global pioneering design. What do you expect for the future design?
5只猫：请你们对未来全球先锋设计的发展趋势作出解读，你们对未来设计的期许是怎样的？

MI5: Before this crisis we expect a lot of imagination and creativity to re-invent the world.
MI5：在这次危机之前，我们期待能用丰富的想像力和创造力重新"创造"这个世界。

COLOR REVOLUTION
色彩革命

Product Name : New Youth Centre In Rivas Vaciamadrid
Architects : MI5 Arquitectos (Manuel Collado Arpia y Nacho Martín Asunción)
Ingenieer : Juan Travesí (Estructuras)
Architect Company : Dragados
Contractor : Eider Holgado, Richar y Diego Barajas.
Location : Rivas-Vaciamadrid (Madrid, España)
Design Date : 2006
Finished Building : 2009.11
Area : 2244m²
Photographer : Miguel de Guzmán /www.imagensubliminal.com

项目名称：巴西亚马德里新青年中心
建筑师：MI5 建筑师（曼努埃尔•哈比和纳乔•亚松森）
工程师：胡安•特里维斯（建筑结构）
建筑公司：德哈加多斯
合同商：艾道•郝加多、理查德和迭戈巴拉哈斯
地址：里瓦斯 - 巴西亚马德里（西班牙马德里）
设计时间：2006 年
交工时间：2009.11
面积：2244m²
摄影师：米格尔德古斯曼

From the beginning, the project was conceived as the possibility of making the "underground" visible, a construction devised as a radical manifestation of Madrid's outskirts youthful spirit in general, and of Rivas' youth groups in particular.

The project aspires to become an explicit "teen" communication vehicle by appropriating their language and their voices as the ingredients of the project. In this way, the project's team embraces all Rivas's youth groups by means of an open participation process, in which the future users of the centre, combined with technicians and politicians, will contribute their decisions, their concerns, their fantasies and their aesthetics to create a contemporary "social monument".

The end result of this process is a public structure with a punk spirit, intensely burdened with content and articulated around programmatic centres conceived as activity explosions, which are erected as meeting and exchange points of the emerging communities.

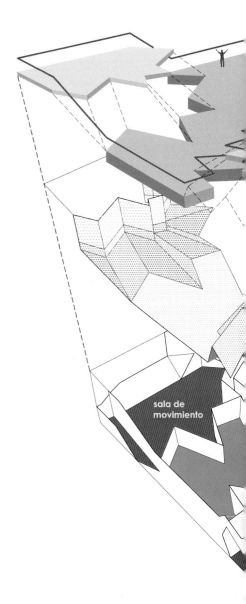

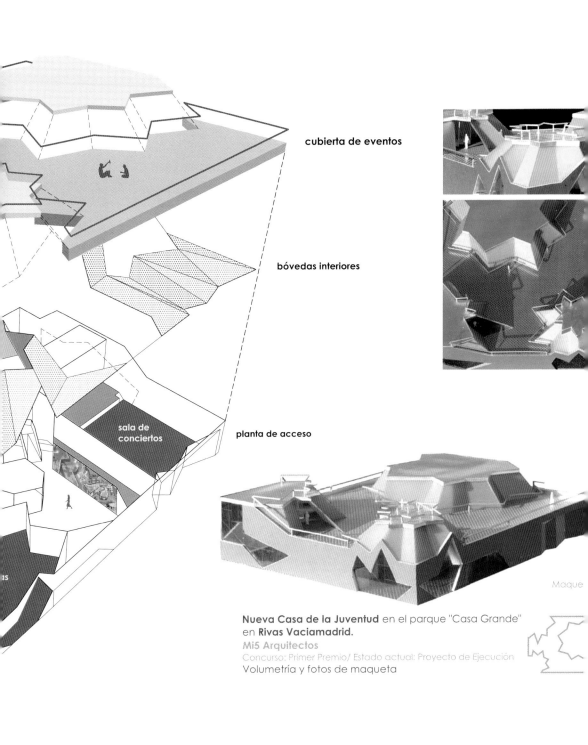

cubierta de eventos

bóvedas interiores

sala de conciertos

planta de acceso

Nueva Casa de la Juventud en el parque "Casa Grande" en **Rivas Vaciamadrid**.
Mi5 Arquitectos
Concurso: Primer Premio/ Estado actual: Proyecto de Ejecución
Volumetría y fotos de maqueta

最近落成的青年中心项目出自西班牙工作室 MI5 旗下的建筑师之手，它位于马德里郊区的里瓦斯。为了迎合年轻人的思维想法，我们参考了马德里的郊区文化，运用涂鸦和漫画手法将亮丽的色彩运用到整座建筑中。这个青年中心在开放的大型空间内设置了座位区，以便当地青年举行会议和交流活动。我们在建筑的角落开辟出星形空间，旨在表现该建筑的内部爆发性。

MI5 崇尚将幻想作为城市进程中最强大的动力源泉，里瓦斯青少年中心是一座单层建筑，主建筑体是一个长方形的盒子，然而我们没有让它像盒子那样"中规中矩"，而是在它的表面打造出若干个多边形的钢管造型，并使其"浮"在建筑体上。

最开始，我们的用色方案深受建筑气息感染，参考了城郊文化的用色体系。彩色元素借助涂鸦和连环画，来表现三维的居住世界。对比鲜明的图片和色彩，我们称之为"砰！"。同时MI5 在设计里瓦斯青少年中心的时候，将涂鸦这一艺术形式从平面上挪到了空间中，用明亮的色彩和夸张的造型打造出了一座如同"立体涂鸦"一般的建筑。

这个项目适当的运用青少年的语言和意见作为其中的元素，希望能成为一个特色鲜明的"青少年"交流媒介。MI5 通过一个开放分享的流程深入里瓦斯所有的青少年团体和他们交流，在这个过程中，中心未来的使用者与技术专家和政治家一起，贡献出他们的判断、他们的关注点、他们的梦想以及他们的审美，来打造这座当代的"社交典范"。

METEOR DIAMOND
流星宝石

Product Name : Jouth House In The Old Fuenlabrada Town Hall. Competition Project
Architects : MI5 Arquitectos (Manuel Collado Arpia y Nacho Martín Asunción)
项目名称：弗拉布拉达大会堂旧址竞争项目中青年中心
建筑师：MI5 建筑师（曼努埃尔•哈比和纳乔•亚松森）

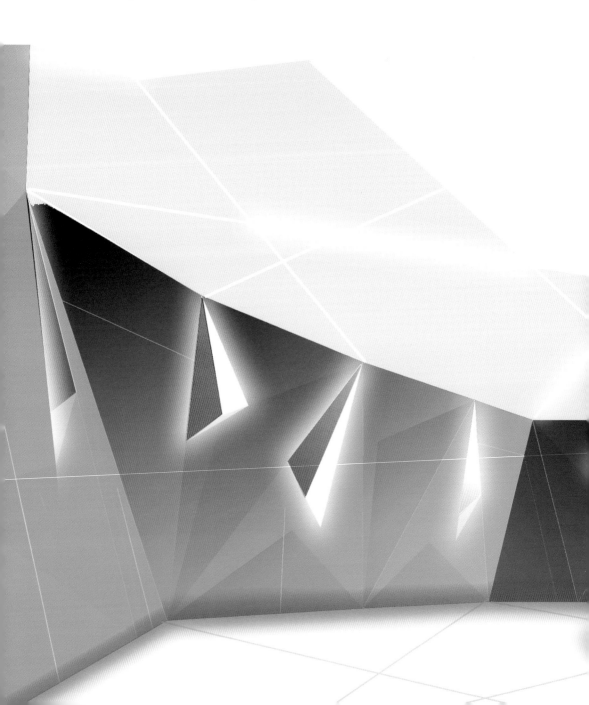

Recycling old public buildings to adapt them to new programmes attain here their maximum expression. The objective is to transform the old Town Hall, with all its municipal areas, in the new Youth House. This entails transforming an inherited, rigid area, therefore conflicting with what a youth department programme in Madrid's outskirts needs.

这里最大化地体现了"旧貌换新颜"的理念,旨在将大会堂旧址及其所有市政区域改造为新的青年中心。改造涉及的区域原本是十分僵化的,与马德里郊区的青年项目需求有所冲突。

There were two simultaneous and obvious difficulties to surmount. First, overcoming the resistance such a representative building as the old Town Hall embodies, when having to construct a clear identity for programmes created by youth groups. Second, the building's inadequate geometrical configuration, inappropriate for the streamlined development the new activities require.

弗拉布拉达大会堂旧址青年中心竞争项目。为了达到这一目标,我们要跨越两座"大山":首先是要树立新的身份,打破建筑原本作为大会堂的身份,让青年中心成为新的主角;其次就是,原先的建筑缺乏几何建构,不适合新活动的现代化要求。

In the face of this situation, we decided to intensify and reconstruct through a tuning process, by creating an identity which would allow the new users to appropriate themselves of the building. In this way, we have inserted/ embedded a large piece in the shape of a meteoric rock or "mutated diamond" en the building's axis. Thanks to it, we reshape the relationship with the area in its section and we restructure the floors so that all the activity revolves around this piece.

在这种情况下,为了让使用者找到对青年中心的认同感,我们在大楼的轴线上插入了或者说植入了一大块形似流星石或者"突变宝石"的设计。因为这个设计,我们重新确立了空间关系,从而重建了地板,使所有的改造思想都可以围绕着这个设计展开。

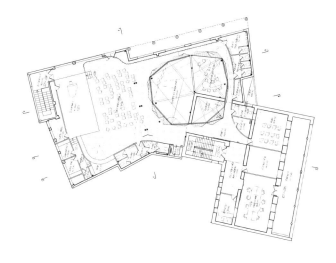

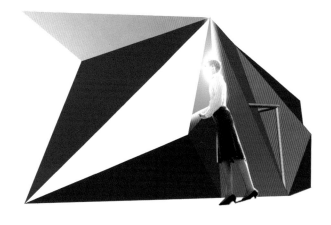

SUN CITY
阳光之城

Product Name : Slovenia. Solar Housing
Architects : MI5 Arquitectos (Manuel Collado Arpia y Nacho Martín Asunción)
项目名称：斯洛文尼亚•太阳能房屋
建筑师：MI5 建筑师（曼努埃尔•哈比和纳乔•亚松森）

First of all we must understand that Izola is the sunniest place in Slovenia (with 2500 hours of sun a year), reason for which retaining and exploitation this natural resource becomes indispensable.

The strategy developed arises from the redefinition of a solar orchard (photovoltaic solar plant) and leads us to a new binomial, fruit of combining the typical Mediterranean landscape and the solar garden.

首先我们必须了解，伊佐拉是斯洛文尼亚日照最丰富的城市（每年有 2500 小时的日照），正因为此，保存和利用自然资源变得十分必要。

我们的策略源于对太阳能果园（光电池太阳能工厂）重新定义的思考，同时，在策略的引导下，我们将两大要素——典型的地中海景观和太阳能花园结合起来。

Usually this blend between generated solar energy and the natural environment is limited to the simple consumption of surface in those places with high radiation ratios.
将系统产生的太阳能和自然环境如此融合，往往受到上述高辐照比地区仅仅是地表消耗的限制。

The challenge in Izola is combining a tourist resort with a solar plant, where the energy surplus is used for what is commonly known as 'charging the batteries', that is, for all those leisure activities which can benefit from this clean energy.
伊佐拉项目的一大难题是将观光旅游和太阳能工厂的很好地结合在一起，过剩能源被用于众所周知的"电池充电"活动，也就是说，所有休闲活动都可以受惠于这项清洁能源。

Thus we coin a new term: solar hedonism, as the state in which we can pleasantly enjoy all the possibilities the sun's exploitation has to offer.
因此我们创造了一个全新术语：太阳享乐主义，以表达我们可以尽情享用太阳提供给我们的能量的一种状态。

Alike plants, these receptive and retaining units grow in search of light on the most typical Mediterranean landscape fragments, which we have decided to christen 'Mediterranean camouflage pattern'.
如植物一样，这些日光接收和储存设备在最为典型的地中海景观片段中追随日光生长，我们认为这情很符合所谓的"地中海模式"。

Lastly, a large plant icon shelters the groups and boldly characterizes the outline of this new Mediterranean tourist settling.
最后，巨大的工厂标志遮住了设备组，彰显了这次全新的地中海观光场所的轮廓特征。

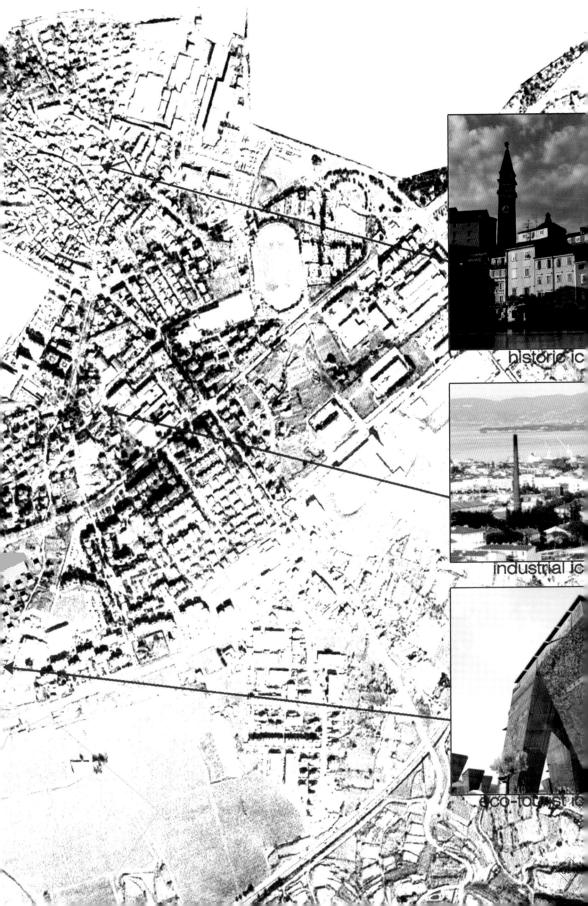

historic ic

industrial ic

eco-tourist ic

ROCK PARTY

礁石乐园

Product Name : Maritime And Sport Park Of Azaña-Acoran, Tenerife
Architects : MI5 Arquitectos (Manuel Collado Arpia y Nacho Martín Asunción)
项目名称：特内里费岛阿桑娜 - 阿科兰海上运动乐园
建筑师：MI5 建筑师（曼努埃尔•哈比和纳乔•亚松森）

apliocación del pavimento en el área de a[...]

WHY AN ARTIFICAL REEF?
为什么要建一处人工礁石？

Aware of the risk of developing urban centres near the coastline, we propose the construction of a new earth-sea ecosystem which will incorporate in its definition the life of the inhabitants of the natural environment within the new programmes associated to a new maritime sports park. We chose the Reef as a conceptual model of intervention, such as this is understood as an exemplary element of coastal protection and how it can restructure new habitats, apart from its landscape value. We suggest the construction of an artificial reef through the following work tools:

意识到在海岸线附近建城市活动中心的风险后，我们建议建一个全新的地球 - 海洋生态系统，在全新海上乐园项目框架内的自然环境中定义居住者的生活。我们选择礁石作为介入的概念模型，除了具备一定的景观价值之外，还可以将其理解为海岸保护、重建新栖息地的代表元素。我们建议通过下列工具完成人工礁石的建筑工作：

-Prefabricated Reef Ball pieces.
Use concrete prefabricated elements (with reusable formwork) available in the market, previously used and tested for the construction of artificial reefs.
- 预制海礁球片
采用预制混凝土要素（带可重复利用模板），这些要素可在市面上买到，属于二次利用，且通过了人工礁施工测试。

APLICACIÓN TERRESTRE

como elemento ornamental y mobiliario del parque (soporte floral, iluminación, protección contra viento, juego de niños...)

APLICACIÓN MARINA

como formación de barrera protectora de playa artificial y generador de ecosistema submarino

-Surfaces treatment
With the intention of regaining the experience of travelling along a reef, horizontal surfaces will be put in place, based on a model of suggestive seabed photographs. This transfer will be destined to material, texture and colour definition within the different park's areas (lanes, gardens, squares, bathroom areas…) which will result in a pedestrian scenery intimately interwoven with the underwater landscape.
- 表面布局
为了找到在礁石观光旅游的感觉，水平表面将以一组海床意向图片为基础进行摆放。同时，必须根据乐园的不同区域（巷子、花园、广场、浴室区域......）做出不同调整，选用不同的材料、质地和颜色，最终形成与水下景观紧密交织的步行风光。

 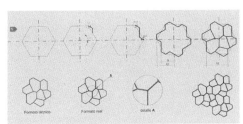 

geometrización para formación de pavimentos aparentemente irregulares

fotografía en detalle de formaciones coralinas

DESIGN JOY
乐活设计

Product Name : Youth Center, Getafe
Architects : MI5 Arquitectos (Manuel Collado Arpia y Nacho Martín Asunción)
项目名称：赫塔菲青年中心
建筑师：MI5 建筑师 （曼努埃尔•哈比和纳乔•亚松森）

1- DECENTRALIZING JUVENILE ACTIVITY: 丰富青少年活动区域
The creation of a new appealing activity space addressed to the younger generation is proposed as an alternative to the one in the capital's centre, capable of becoming the focal point of the new emerging South.
新的青年中心专为年青一代设计，是青少年向往的集散地，将与首都市中心的青年中心齐名，成为南部新兴区域的一大亮点。

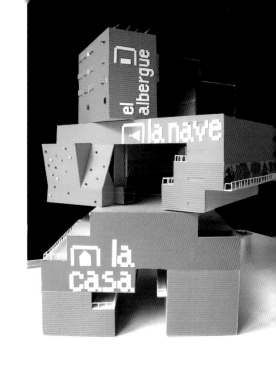

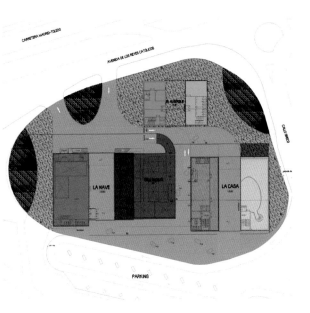

2- ORGANIZING ACTIVITIES 组织各类活动

A programme will be established, conceived as open and flexible around three principal nucleus (the house, the nave and the youth hostel), but which can be spread out from them, in an attempt at a recreational and cultural colonization of the environment.

其中一个项目将围绕三大主要核心（房子、中殿和青年旅馆）开放、灵活地展开，但也可以分散开来，以创造一个以娱乐和文化主导的环境。

3- PROGRAMMING 3 WELCOME THRESHOLDS 设三个迎宾入口

The formalization of the activity nucleus will be undertaken with the main objective of defining hyper significant elements which induce and represent the desired welcoming and approachable character.

确立活动核心的过程，以定义鲜明要素为主要目标，其中包括引入和表现众望所归、形象亲民的迎宾人物。

BLUE GLAMOUR
蓝色飞扬

Product Name : Youth Centre In The Old Fuenlabrada Town Hall
Architects : MI5 Arquitectos. Manuel Collado Arpia y Nacho Martín Asunción
Engeneering : Maricarmen Nombela
Contractor : Ciopsa
Photographer : Javier de Paz
项目名称：福拉布拉达大会堂旧址青年中心
建筑师：MI5 建筑师（曼努埃尔•哈比和纳乔•亚松森）
工程师：马西卡曼•诺贝拉
合同商：希皮萨公司
摄影师：贾维尔•帕兹

西班牙 MI5 建筑事务所的建筑设计师对西班牙青年活动中心的室内设计进行了改造。设计师表现了新一代青少年的朝气与创造能量。意图将这里变为青少年朋友之间沟通的工具，而不仅仅是一个活动场所。设计过程中，活动中心未来的主人们——当地青少年受邀参与其中，描绘出他们理想的使用状态。在设计师和青少年们的共同创意下，一个新的设计诞生了。

Emptying the old Fuenlabrada Town Hall of the elements built inside it is the first step of the transformation process of the execution project established for this area. As a computer formatted integral process, the area is reduced to its basic structural elements, allowing an episodic, programmatic construction generated by the different and changing uses of the space.
在该地区落实该项目需要完成一系列改造，而改造的第一步就是将福拉布拉达市政厅旧址内的要素清空。通过电脑整合分析，空间被简化成基本的建筑要素，让我们可以分阶段、按计划地根据空间功能的不同和变化来施工。

Representing such architecture, invisible but fraught with forces which pierce and reconfigure it, is the main objective of creating the centre's identity. Intervention concentrates on the construction of an immersed area, activated by the qualified presence of the installations which go hand in hand with the users' activities throughout the whole building.
确立青年中心的身份感主要是为了彰显此类建筑"力透纸背"的无形力量。整个建筑内，高质量的设施和用户的活动紧密关联，展现了介入设计的重点，那就是尽享其中。

WON-MYOUNG HYUN   Pioneering Design Can Be Expected

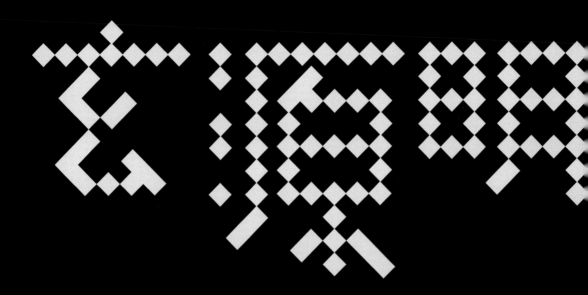

先驱设计未来可期

WON-MYOUNG HYUN
玄源明

Cheongju University, Design Industry
Kookmin University, Design graduate school , interior design
Present Inexdesign
Kyungwon University , interior architecture, adjunct professor
KOSID(Korea society of interior architects/ designers) regular member
KIID (Korean institute of interior design) / manage director
The 6th Modern Decoration Media Prize 2008 , grand prize

毕业于清州大学工业设计系和国民大学设计研究生院室内设计专业
现任依尼克斯设计公司设计师
兼暻园大学室内建筑系副教授
韩国室内建筑师／设计师协会常任会员
韩国室内设计研究所管理主任
曾获2008第六届现代装饰传媒奖大奖

INTERVIEW: WON-MYOUNG HYUN
玄源明访谈

5Cats: As our neighboring country, what do you think is the difference between design culture in ROK and that in China? What exchanges can we make in interior design?
5只猫 作为邻邦,您认为韩国与中国的设计文化有什么区别? 在室内设计方面可以开展哪些交流呢?

Won-Myoung Hyun: Answering to this question is very difficult to me. Thus, I would like to give my general opinion. Design cultures in both countries reflect their own social and cultural differences. The two countries have big differenced in geography and climate and that has led to the social and cultural differences between them. Naturally those social and cultural differences strongly influence the development of their identities and hence their design cultures.
玄源明:这个问题对我来说比较难回答。我可以说说我的大致观点:两国的设计文化都体现了各自的社会和文化特点;两国在地理和气候等方面都有着很大差别,因此造成了社会和文化方面的差异。很自然地,这些社会和文化的差异深深地影响了他们文化认同的发展以及他们的设计文化。

The interior designers in both countries should discuss and exchange their own design cultures, not their popular design trends, based on their mutual respect on the different cultures. One possible way is to have regular exchange exhibitions between the designers from the two countries, which may increase mutual understanding and thus be mutually beneficial.
玄源明:两国的室内设计师应该在对不同文化相互尊重的基础上探讨和交流各自的设计文化,而不是本国的流行设计趋势。我们可以通过在两国间举行定期展览的方式进行交流,增进相互理解,使双方互惠。

5Cats: How do you balance national cultural identity and international impact in your work? What influence do national cultural characteristics have on your design concept?
5只猫:您在设计作品时如何平衡本国文化特征与国际化的碰撞?本国文化特征对您的设计理念产生了怎样的影响?

Won-Myoung Hyun: It is natural that my design works have been affected strongly by my country's cultural identity because I have been raised and working in Korea. I think, all aspects of Koreans' cultural identity are represented unconsciously in my design works. Thus, to balance between my country's cultural identity and international impact, I try to partially adopt some interesting concepts from other cultures in my design works using global languages, rather than special languages. Therefore, my design works mainly contain Koreans' cultural identity but they also contain a certain degree of other countries' cultural identities. Regardless of the cultural identities, each of our design works is expressed as a whole in a global language and thus everybody in the world may easily understand them.

玄源明:我的设计作品深深地受到我的国家文化认同的影响,这很自然,因为我是土生土长的韩国人。我认为,韩国的文化认同潜移默化地影响着我的设计作品。因此,为了平衡文化认同和国际影响,我试着用国际语言,而非专属语言来诠释我的设计作品,并从其他文化中汲取一些有趣的想法。因此,我的设计作品中除主要体现了对韩国文化的认同感外,一定程度上也包含了对其他国家的文化认同。抛开文化认同不讲,就整体而言,我的每一件设计作品都是通过国际语言表达的,这样世界上的每一个人都能很容易地理解我的作品。

5Cats: Under "globalization", how can Asian or Korean culture and life style keep its own unique personality?

5只猫:在"全球化"的背景下,您认为亚洲,韩国的文化及生活方式应该如何保持自己独特的个性?

Won-Myoung Hyun: I can firmly say "Yes". Asia's outstanding cultural competitiveness has been well recognized in the world. Continuous efforts to globalize Asia's characteristic cultures and life styles, while maintaining their unique identities, may strongly affect the other world without loosing their unique personalities under globalization.

玄源明: 我可以肯定地说"是的"。亚洲杰出的文化竞争性得到了世界的公认。不断努力将具有亚洲特色的文化和生活方式全球化,并维持他们独特的特色,将在全球化不丧失自我的同时对世界上的其他国家造成深远的影响。

5Cats: Cobb West once said that if the owner does not like a building, then there is no need to explain. It must be the fault of the architect. How do you deal with the relationship with the owner in Korea?

5只猫：柯布西埃曾经说过，如果业主不喜欢某建筑，那么无须去辩解，肯定是建筑师错了。在韩国，您是怎么处理和业主之间的关系的？

Won-Myoung Hyun: It is a very subtle question. I think, our design works do not exist without our clients. The designers failing to persuade their clients may not be highly recognized and hence they may not continue their design works long time. The failure may come from misunderstanding and/or neglecting their clients' needs and circumstances. I think, the best deign work starts from understanding and reflecting clients' needs and circumstances and that may eventually lead to the highest level of satisfaction to the clients.

玄源明：这个问题很微妙。我认为，如果没有客户，那么我们的设计作品也就不复存在了。无法说服客户的设计师就无法得到较高的认同，而他们的设计作品也不会持久。这个问题可能是由于误解和（或）忽视了客户的需求和条件造成的。我认为，最好的设计作品始于对客户需求和条件的理解和反映，而最终也会使客户达到最高程度的满意。

5Cats: Some people say Korean interior design is more rural in style and has strong flavor of home living, similar to Chinese decoration, while it also shares some thoughts with the Western style. It takes advantages of many others. How do you see Korean interior design?

5只猫：有人说，韩国室内设计风格比较注重田园风格，居家气氛浓厚，与中式装修十分类似，但是装潢设计理念又有很多西方的痕迹，可以说是取百家之长。您怎么看待韩国的室内设计的呢？

Won-Myoung Hyun: It is extremely difficult to define current Korean interior design in simple words because there are so many outstanding designers with their own styles. Previously many Korean interior design works were mainly based on rural styles while taking some advantages of Western styles. Many designers in Korea just followed such trends at that time. Now Korean interior design is in a transition stage. Nowadays many interior designers in Korea try to clarify their unique identities and to reflect the spirit of the time.

玄源明：要用简单的几句话来概括当前的韩国室内设计实在是太难了，因为韩国有很多有自己独特风格的杰出设计师。早前很多韩国的室内设计作品主要依托于乡村风格并吸收了西方建筑风格的一些优点，当时很多韩国设计师都是跟着这股风潮。现在，韩国的室内设计正在转型。今天，很多韩国室内设计师试着诠释他们自己的特点并反映时代的精神。

5Cats: How do you think can design satisfy men's life? What is your ideal "home" like that is designed by yourself?

5只猫：您认为设计应该如何才能满足人的生活？您理想中，给自己设计的"家"应该是什么？

Won-Myoung Hyun: Surely, excellent design works can satisfy human life. I think, it is the era of women and hence men feel satisfaction to the decisions made by women. I have limited numbers of home design works, but I think an ideal home is not a simple place for relaxation. It should be a new type of creative place functioning as both a community space and an independent space for family members.

玄源明：当然了，出色的设计作品能够满足人们的生活。我认为，这是一个女性的时代，男性满足听从于女性做出的决定。我的家居设计作品有限，但我认为我理想的家应该不仅仅是一个放松的地方。它应该是一种具有创新功能的新型家居房屋，既有社交功能，又能为家庭成员提供独立空间。

5Cats: In your eyes, which of your projects can best represent you and your design style?

5只猫：在您的眼中，自己作品中的哪个项目最具代表性，最能够充分展示你的设计风格？

Won-Myoung Hyun: My design style changes with time, but the hip-hop multi-shop 'paper3' may represent best my design style. The paper3 was introduced previously in China.

玄源明：我的设计风格与时俱进，但嘻哈多元店"纸3"可能最能代表我的设计风格。"纸3"也曾在中国做过宣传。

5Cats: What are the most considered issues in real projects based on years of design practice?

5只猫：经过多年的设计生涯实践，您认为设计师在实际项目中最应该考虑的几个问题是什么？

Won-Myoung Hyun: There could be many issues, but I devote mainly on metaphor and story when I work on my projects.

玄源明：可能有很多问题，但我在设计我的作品时主要看中的是隐喻和故事性。

5Cats: Which cities in Asia have you been to? How do you like those cities? What influence do exchanges with architecture designs in other countries have on you?

5只猫：您去过亚洲的哪些城市，感觉如何？在这些与其他国家的建筑设计交流中，对您产生了哪些影响？

Won-Myoung Hyun: Beijing, Shanghai, Shenzhen, Tokyo, Osaka, ...
Every city has its own characteristic nature in addition to some common things. In this regard, I like all the cities visited. Although I have some interest in fancy downtown

area in a city, my main interest is in the alley and the market outside the downtown area. The downtown area has been built in a kind of standardized criteria whereas the alley and the market preserve well the history of the city and the way of general people's living. This, I value the architecture designs in those areas high and try to find universal languages to creatively include inspirations from the architecture in my design works.

玄源明：北京、上海、深圳、东京、大阪……每座城市除了有共通点外，都有自己的特色。从这个角度看，这些城市我都喜欢。尽管我对城市的市中心繁华地带也有兴趣，但我的主要兴趣点还是在市中心以外的小巷和市集。市中心地带一般都是标准化的建筑，而小巷和市集完好地保存了城市的历史和普通人的生活方式。所以，我更看中这些地方的建筑设计，并试图寻找国际性的语言，从建筑上汲取创新灵感并将其应用到我的作品中来。

5Cats: When you look into the future, what do you think of the trend of design? What kind of design can be still called pioneering design in the future?

5只猫：展望未来，您觉得设计发展趋势是怎样的？什么样的设计能在未来依然称之为先锋设计呢？

Won-Myoung Hyun: I think, a pioneering design in the future is the design with intrinsic sensitivity rather than that focusing on superficial or morphological concepts. A pioneering design changes and evolves with time, but it should be the one which stimulates emotions deep inside human beings and makes people keep thinking of it. This kind of design may prosper in future.

玄源明：我认为未来的先驱设计是那种能给人带来内在感受的作品，而不是那些只注重外在和形态概念的作品。先驱设计随着时间的变化而变化和演进，但应该能够震撼人心、引人深思。这种设计将在未来大有发展。

CLEAR INKY BLACK
纯净的墨黑

Project Name : Books Cooks (Food Gallery)
Design : Won-myoung Hyun/Inexdesign
Operator : Inexdesign
Function : Food Gallery
Location : 177-4,Gahoe-dong Jongno-gu Seoul,korea
Area : 157.5m²
Design Date : 2010.3
Construction : 2010.5
Flooring : Epoxy,Boron
Wall : Korea Paper,Contour Picture drawn in India Ink,Printed Image,Steel Plate
Ceiling : Korea Paper,Fabric,Electric Tent

项目名称：书厨（餐厅）
设计师：玄源明 / 依尼克斯设计公司
运营方：依尼克斯设计公司
功能：餐厅
位置：韩国首尔嘉会洞钟路区 177-4 号
面积：157.5m²
设计时间：2010.3
工期：2010.5
地板：环氧、硼
墙壁：韩国纸、印度墨轮廓画、打印图像、钢板
天棚：韩国纸、纤维、电动幕

As you can know from its name "Books Cooks", this is a commercial space to introduce and sell food, drinks, and beautiful tableware collected by clients, as well as various books related to cooking, aiming to incorporate the function of a commercial space ane the individually of Books Cooks only without damaging the character of a Korean -style house , it has an open kitchen table at the center of the courtyard, reminding visitors of a 'NANTA' performance venue. This intends to express dynamic factors enabling performance in a very static space of Korea-style house. Encircling the center of the courtyard(stage), the hall symbolizes seats. Spectators enjoy the food making and serving performance, as if they viewed a performance, while sitting on a stylobate naturally.

书厨，顾名思义，是推介和销售食物、饮品和客户收集的精美餐具以及各种烹饪书籍的商业化场所，它集各种商业功能于一身，在不影响韩式建筑风格的同时体现了餐厅的独特品味。院子正中间是一个开放式的厨房桌台，让顾客想起韩国的"乱打秀"（译者注：一种利用煎锅、汤勺、刀等厨具再加上韩国的传统音乐、实物游戏以及幽默故事组成的表演）。在静止的韩式房屋中创造出动感的元素，为表演制造了空间。环绕院子（舞台）中央的大厅象征座椅。观众可以自然地坐在高柱椅上欣赏食物制作和端出的整个过程，就像欣赏一场表演。

All elements here are inky black without color. Furniture and utensils were fabricated of tinplates and inky blacks only to incorporate inky blacks as they are, by spraying Indian ink on the Korean paper link throwing it forth. The color in the space is filled with foods, small pieces and people, and the only color exists in the object, on the back of the counter, which was printed several thousand times like modification.

这里所有的元素都是纯墨色的，不掺杂任何其他颜色。家具和餐具都采用漆黑的镀锡铁皮作为材料，用印度墨喷在韩国纸上，并向前抛掷，形成浑然的黑色。空间里的颜色满是食物、细小的片段和人群，而物体上唯一的颜色藏在柜台后面，而上千次的涂抹为其形成了独特的装饰。

TIME TRAVEL
时光穿越

Project Name：Hangarak (noodle shop)
Design：Won-myoung Hyun/Inexdesign
Function：Noodle Shop
Location：268-4, Seohyun-dong Bundang-gu Seongnam Gyeonggi-do,korea
Area：26m²

项目名称：Hangarak（面馆）
设计师：玄源明 / 依尼克斯设计公司
功能：面馆
位置：韩国城南京畿道洞盆唐区徐洙洞 268-4 号
面积：26m²

Jang, a traditional market opening every five days in Korea, opened in the bustling town. There is turmoil among sellers and customers all in white. There are excited children and shy ladies and they are full of curiosity and joy. Everywhere in the market place there are unpacked bundles which were brought by the peddlers for today's sale. In a side of the Jang, the smell of anchovy broth in a huge cauldron with rising steam invites hungry people.

"张"是韩国每五天开一次的传统集市，位于城里的繁华地段。白天里，市集上的店主和顾客争执不休。孩子们欢笑跳跃，妇女们羞涩内敛，充满好奇和喜悦。市场上到处都是当日小贩们带来的、要在市场上卖的成包货品。在市场的另一端，大锅里热气腾腾的凤尾鱼汤让人垂涎欲滴。

Tak! Tak! Tak! Song! Song! Song!Huruk! Huruk!, ... (Korean imitative words expressing very fast and delicious eating of noodles)
呼噜噜，呼噜噜，顾客们狼吞虎咽地吃着热面条。

The space in the restaurant is divided into a cooking space and a hospitality space. The design of the restaurant focused on operational efficiency as well as open space feel. Between the two different spaces,separated by a crudely designed table, cooks and guests breathe together, exchange words of blessing each other,and share their life stories. The restaurant serves noodles with full of taste and affection to its guests and thus it serves as a space where its guests feel escaping from their busy daily life and recollect their happy memories in their childhood.

餐馆内部被分割成烹饪区和用餐区。餐馆的设计追求使用效率和开放的空间感。一张粗修的桌子将两个区域划分开来，厨师和客人呼吸同一片空气，相互交换友善的话语，分享生活的故事。餐厅为客人准备了充满香气和情感的面食，使得到此用餐的客人们似乎逃离了日常的繁忙生活，重拾儿时的幸福记忆。

To express simplicity and blandness of noodles, my design team

tried to pack full of noodles in the space and avoided unnecessary finishes and decorations. Like kneading dough elaborately and making noodles from the dough, this objet was made through careful weaving of our staffs' elaborate efforts with an aim that the objet explains the space well and delivers strange delight to the guests.

为了表现面馆的简约和亲和力，我的设计团队试着在空间内填充面条的画面，又避免显得过于矫饰。就像精心揉捏的面团和从面团中拉出的面条，我们的作品也交织着团队成员的良苦用心，为顾客营造出良好的氛围和别样的风情。

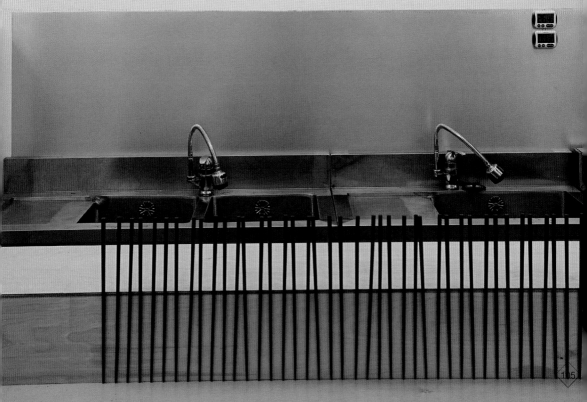

먹거리차림
꼽빼기전원추가

멸치국수 사천원
= 🌶+🥒+🍅+🥚+🧄 ≡🍜

비빔국수 사천오백원
= 🌶+🥒+🍅+🥚+🧄 ≡🍜

한 이란 크다는 뜻을 내포한
우리네 손수를 입니다.
매일 직접 뽑은 면과 국내산
천연재료만을 사용하여 옛맛을
찾아오신 분들께 맛깔난 국수와
우리네 정을 푸짐히
담아드리고자 하는 마음을
담았습니다.
언제나 정겨운 우리네 정성과 같이
한 그릇 한 그릇에 정성을 다하는
한 가락이 되겠습니다.

믿을수있는 먹거리
한가락

MARY JANE

玛丽·珍

Project Name : Mary Jane(fashion shop)
Design : Won-myoung Hyun/Inexdesign
Function : Fashion Shop
Location : 454-1,Shinbu-dong Chonan Chungcheongnam-do,korea
Area : 1F-51m², 2F-51m²

项目名称：玛丽·珍（潮店）
设计师：玄源明 / 依尼克斯设计公司
功能：潮店
位置：韩国忠清南道天安新娘洞454-1号
面积：一层51m²；二层51m²

'Swan Lake' … Odette of a lofty gesture! This classic talks at thesame time about beauty and magnificent love through the ballet whichis the greatest language expressed by a human body. An ideology ofballet, to create the existence that transcends the body of human!Isn't it roman of women who crave for beauty rather than reality?Façade of the building symbolizes ballerinas acting a swan. Thecentral object in the shop is the queen Odette and an essence ofbeauty women craving for. It is intrinsic beauty of women, which iselegant, pure, and lofty.This shop is a bonded shop selling specially women's wear andaccessory. A clothing department is located at the first floor; thesecond floor houses recess space and an accessory corner. I tried tominimize expression except the least functions. The central object isa fitting room and other rooms were designed to be disguised bycurtains.

"天鹅湖"奥德特的倩影！芭蕾舞展现的美和伟大的爱情是人类肢体的最有力的语言。用芭蕾的意识形态来创造超越人体的存在！追求精神内涵美的女人难道不浪漫么？店铺的正面是一名天鹅造型的芭蕾舞女演员。店铺的中间是王后奥德特和美丽女人渴求的精神本质：那就是女人的内在美，高贵、纯洁而崇高。这家店铺是女性服饰专营店。一层是服装；二层是休息区和饰品区。除了最基本的功能外，我们尝试将表达尽量简化。中央是一间试衣间和几间帘幕房。

PAPER3
纸 3

Project Name : paper3 (Hiphop Fashion Shop)
Design : Won-myoung Hyun/Inexdesign
Function : Multi Concept Shop
Location : 136-2, Munhwa-dong Chonan-si, Chungnam-do, korea
Area : 171m²

项目名称：纸 3（嘻哈潮店）
设计师：玄源明 / 依尼克斯设计公司
功能：多概念店
位置：韩国忠清南道天安寺文化洞 136-2 号

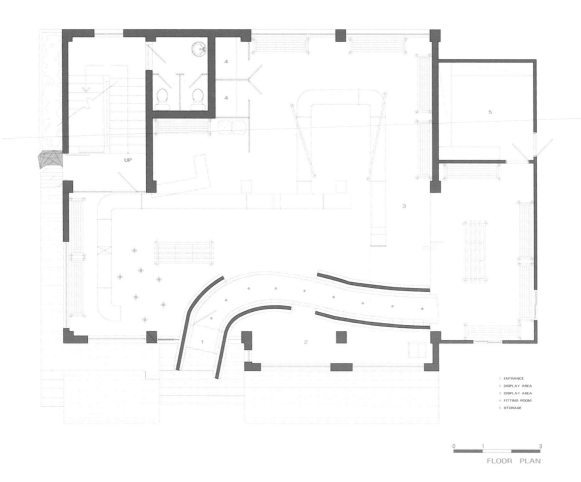

1. ENTRANCE
2. DISPLAY AREA
3. DISPLAY AREA
4. FITTING ROOM
5. STORAGE

FLOOR PLAN

To create a new character, emphasizing on the concept of hip hop, for the desolated building and to give it with an element (a component) of a landmark, I tried to express the language, music, and message of the young generation enjoying the hip hop culture, rather than to focus on the formality and decoration of the shop.

为创造一种全新的特色，强调嘻哈的理念，为荒凉的建筑物增添地标元素，设计师尝试使用音乐的语言，表达年青一代对嘻哈文化的热衷，脱离对正统和装修的专注。

The exterior surface is designed to express well the refined and concise thought and desire, not irrational, resistant, and cynical characters, of the young generation.

外墙要表达的是精巧细致的想法和渴望，以及年轻人非无理的、反抗的和愤世嫉俗的性格。

The approach tube (entrance), passing through the pre-existing architecture mass, destroys the barrier between interior and exterior and also symbolizes the overflowing power and energy of the young generation. This is a warning and satire against hypocrisy of people.
进门的管道（入口）穿过已有的大块建筑，在内外之间破茧而出，象征着年青一代外溢的能量和精力，是对人们虚伪造作的警告和嘲讽。

The interior finish expresses the natural action of the young generation and the movement of B-Boy and also shows well the characters and properties of common materials.
内部的装饰表达了年轻人的自然动作和 B-Boy 的经典舞步，同时淋漓尽致地展现了普通材料的特质和特点。

In this newly designed space, the conflict between the young and old generations vanishes and a message of hope emerges. The clothes and other items sold here are recognized as a merchandise and message of hope to the customers.
在这种全新设计的空间里，新旧两代人之间的冲突消失了，并呈现出一种希望。这里出售的不仅是衣物，还有希望的讯息。

This shop has no official name. Naming the shop was ruled out from the beginning of design. The shop has been hoped to be recognized as a space stimulating imagination and curiosity of the customers. The 'papbar' on the building exterior is a pseudo-name . The exterior text is an musical expression and language for rap.
这家店铺没有正式的名字，给商店取名字在设计之初就被否定，我们希望它能够激发顾客的想像力和好奇心。建筑物外面的"papbar"其实是一个假名，只是一种说唱音乐的表达和语言。

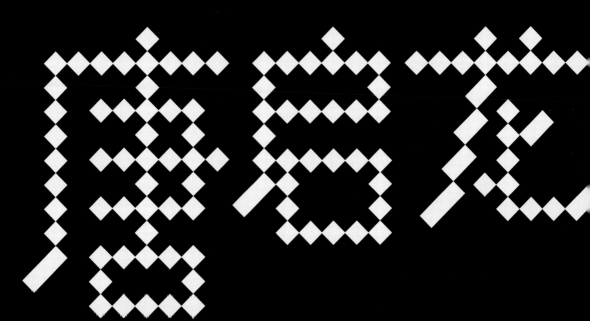

THOMAS DARIEL

Design Is Signature

设计是你的签名

THOMAS DARIEL
唐启龙

关于事务所
Dariel Studio 是总部位于上海的国际室内设计公司。

主要专攻住宅、创意办公室、零售概念店和高端服务业（酒店、精品酒店、水疗、俱乐部、酒吧和餐厅）的创意设计。

自从 2006 年，事务所以 Lime 388 为名成立后，先后与众多奢侈品牌和国际集团取得良好的合作。如 Christofle, Hermès(爱马仕), MHD, Starwood(喜达屋), Golden Tulip(金郁金香酒店集团)等。

Dariel Studio 同时也和来自澳洲的希腊籍名厨 David Laris 合作餐饮概念的设计，并在上海开设了近 20 家餐厅。近 5 年来，事务所在上海，周庄，北京等地完成 50 多处室内设计项目。

2011 年是蓬勃发展的一年，几个大项目的顺利完成并且被国内外媒体广泛转载，使得事务所荣获年度最具潜力设计师大奖。

关于设计师
Thomas Dariel 是这家事务所的主案设计师及创始人。

作为法国家具设计师的曾孙，爵士音乐家的孙子，以及建筑师的儿子，他从其家族中继承了力量和灵感。

通过学习和不断增加的阅历，这位从巴黎来的设计师在 2006 年来到中国，开创了其事务所——Lime 388。

Lime 388 在这五年的成功积累后，Thomas 决定将其重新命名为 Dariel Studio，作为其艺术成熟和开拓新目标的里程碑。

相对设计：思考源于实践
结合传统遗产和先锋创新，玩转感觉，混合强烈对比，融合法式设计专长和东方文化影响。Dariel Studio 旨在将现实转化成调动五官感受的体验。

Dariel Studio 通过不断开发新行式，在实践的过程中以其开放的态度展现出了超常的远见和潜力，并且总是奉行规则和客户的需求标准。以这种方式创造室内设计，一种新的表达方式就此诞生，在每个项目上都体现出由激情驱动而颠覆传统和一成不变的特色。

大气，优雅，活力，专业，创新，并且高度理解客户的需求使得我们的设计作品达到最高的水平和质量。

奖项和提名
2010: 第八届现代装饰国际传媒奖：
年度最具潜力设计师大奖
年度办公空间大奖 (Imagine China 办公室项目)

2011: 金堂奖 2011 China-Designer 中国室内设计年度评选：
年度优秀餐饮空间设计作品（YUCCA 酒吧项目）
International Arch of Europe Award (IAE) – 金奖

2012:The Restaurant & Bar Design Award 的提名
The Andrew Martin International Award 的 3 个项目提名

9300KM,
DREAM IS TOUCHABLE
9300 公里，梦想零距离

9300 公里，这是上海与巴黎的距离。然而，这位年轻巴黎设计师与他的梦想之地上海之间，却没有距离。

当其他 23 岁的青年都在享受生活时，Thomas Dariel 已经在远离故乡的城市开始了他们的事业。他们的设计相对于他的年纪来说显得异常成熟。起初，人们一度怀疑在如此年轻的时候创办一家设计公司是否可行，但是他运用自己的天赋消除了所有疑虑并证明了他的与众不同。

如今，已经第 5 年了。在这 5 年里，他创造了一个个新的传奇：23 岁时只带一件行李来到中国；在同年开办了自己的设计公司；25 岁时与爱马仕合作；27 岁时被评为最有潜力设计师；2010 年与国际名厨 David Laris 强强联手打造完美餐厅......然而传奇从未停止，更多令人惊呼的作品和事迹将接踵而至。

设计感言
设计对于我们来说就是生命，激情和创意是我们的武器，设计师就像导演，要把冷冰冰的空间叙述成鲜活的故事，营造一种意境，而材料和家具就是我们的演员。热衷于将死板的空间做活，热爱挑战。在我们看来，设计就是设计，每一个作品都是独一无二的。设计要根据空间原本的特点考虑美观和功能的结合。法国和中国都是伟大的国家，拥有深厚的文化底蕴，我们喜欢从法国和中国这两种迥异的文化中汲取养分，摄取文化元素，创作出更好的作品。

Thomas Dariel
2006 年 毕业于法国巴黎的 EDNA(L'Ecole de Design Nantes Atlantique)，并获得工业及室内设计国际硕士学位
2006 年创立 DARIEL & ARFEUILLERE–A Lime 388 Company

"他很疯狂！"很多人第一次遇到他都会这么形容。没错，他的确很疯狂：他喜欢在浴缸里读书，每年都搬家，在自家客厅里把古代麻将桌和巨大的石膏像放在一起…他热衷于打破常规并且将生活中各种细节变的充满乐趣。
Thomas 拥有非常积极乐观的生活态度。他一直保持对新鲜事物和想法的好奇心。他喜欢做他自己，每当他与朋友分享他的设计和最新的想法时，他的眼睛总因兴奋而闪烁。他从不假装成为一个学识渊博的伪君子。他独特的个人魅力将人们引领进他的色彩和奢华世界，并流连其中。

作为一个刚拥有 6 个月大宝宝的父亲，Thomas 将其视为新的灵感来源。同时，不断有新的作品问世。没人能预料下一个作品又会是如何另人惊叹。但毫无疑问，肯定是超越无穷的想像力。

INTERVIEW: THOMAS DARIEL
唐启龙访谈

5只猫：提到法国设计，大家都很熟悉，Philippe Starck，保罗安德鲁，地球村让不同国家之间的距离更近了，那么作为一个在上海工作的法国设计团体，你怎么看待中国的设计？

唐启龙：中国设计的发展真是令人印象深刻，在我的印象中，中国5年的发展，在欧洲可能要用20年。2006年我们来到中国时，设计界的抄袭现象十分严重，而今天的中国已经越来越难发现抄袭了，也出现了很多有特色的设计和设计师。如果用欧洲的设计风格比喻的话，中国设计更像巴洛克风格，色彩丰富，中国设计的另一个特色是使用自然材质，木头、石头、大理石等，还有风水等哲学贯穿其中。

5只猫：我们注意到，你们在设计的时候还考虑到了风水因素，我们很好奇作为一个法国设计师团队，是如何将风水这一东方独特的文化纳入你们的设计思维中的？

唐启龙：我们开始工作的时候，会列出工作的项目和要考虑的方面。风水理念在中国很普及，虽然有时候我们会遇到很疯狂的风水大师，但我们非常尊重这种理念，而且常常可以给我们带来新的想法和创意，有需要的时候我们会根据项目的要求将风水理念整合进整体设计中。

5只猫：居住风格和环境是大家关注的事情，设计师会用自己独有的设计方式去处理不同风格的搭配，你们是如何处理新与旧、国与国之间风格的关系的？

唐启龙：创造出自己独特的风格和品位是每一位好设计师追求的目标。很多时候项目本身的特点决定了它适合哪种风格或者说要讲述什么故事。我们会从地址、周边环境、客户需求及空间本身的特点和用途来全方位思考，并不拘泥于某种风格，有时会混搭，但是绝对会创造出独一无二的特色。我们将在周庄完成一个酒店设计的项目，在保持当地水乡文化的原汁原味的同时，融合了部分欧洲元素，很难定义是什么风格，但毫无疑问将是我们的风格，让我们共同期待这个精彩的项目吧。

5只猫：社会在不断向前发展，在你们看来，我们的居住环境在未来将会发生哪些方面的改变？室内设计潮流会向什么方向发展？

唐启龙：幸福舒适将是未来空间和室内设计的趋势，也是人们所共同追求的东西。中国在过去的快速发展中建立了强大的工业，但是现在人们越来越关注幸福感，关注舒适度，追求新鲜有机的食物，洁净的空气、水，互动的、人性化的空间环境，无论是酒店，还是办公室、餐厅，人们所到之处都希望看到有出色创意的、能带给人愉悦和幸福感的空间，人们已经抛弃了塑料和质地差的材料，而越来越多地采用天然的、绿色的、令人们感到愉悦的材料。

5只猫：法国巴黎是著名的时尚之都，世界的时尚潮流都随着法国风向标转动，那么作为法国设计师你们眼中的先锋设计是什么？

唐启龙：我认为，未来的先锋设计将是使用高科技材料制成的人性化的智能服装。在时装设计方面，我们已经有了足够多的图案、款式，你可以看到，大约3个月就会完全变化一次，而今后将更多地关注科技和材料。室内设计和服装设计有很强的关联性，无论从材料、款式、颜色和图案，很快会从服装设计领域延伸到室内设计领域，因此，好的室内设计师一定是掌握服装设计潮流的。

5只猫：近年来，生态建筑的设计理念大量被应用在室内设计之中，请问你们怎么看待这一潮流？在实施环保设计时有什么心得？

唐启龙：我们认为，环保生态设计恐怕已经不能称其为趋势了，在某种程度上它已经是事实了，而且未来必将长期存在。对于环保生态设计，我们首先是尊重然后是赞同，并且已经在我们的很多设计中应用和体现了，我们可以很自豪地说我们是环保生态社区中的一员。

5只猫：资讯的开放给我们带来了极大的方便，但巨大的信息量也的确容易让我们迷失方向，设计师如何在这样的状况下不迷失自己的风格呢？

唐启龙：好的设计师区别于其他人的很重要的一点就是懂得如何甄别好的信息和不好的信息，并且深刻理解不好的信息会严重污染你的大脑。我认为，自己知道如何用自己的理念和逻辑去整合通过甄别的有益信息，并且用以激发灵感，通过思考和原创形成自己的风格。

5只猫：你毕业于巴黎学院，你们坚持的设计理念和观点是什么？谈谈在中国做设计，你们觉得最重要的是注意什么？

唐启龙：创意，跳出固有的框框去思考这是我们认为的设计的灵魂。我将设计看成自己的签名，无论在中国还是在其他任何地方，我都认真工作，认真对待自己的每一个设计，因为设计就是你自己的签名。

5只猫：当你还是一个孩子时，你们希望将来做什么？那时你想成为设计师吗？

唐启龙：对，我从很小就开始对设计感兴趣了，Thomas 从曾祖父到父亲都是设计师，10 岁的时候，他看到了 Philippe Starck 设计的牙刷，并为之震撼，因此立志做一名设计师。

THE FUNKY CHICKEN
疯酷烤鸡

Dariel Studio announce the completion of four unique food and beverage experiences in Shanghai: The David Laris Concepts at Sinan Mansion.
Dariel Studio 在上海宣布四个新的餐饮概念室内设计项目的完工：思南公馆 David Laris 概念餐厅。

The David Laris Concepts at Sinan Mansion are four different venues with inspirations spanning from the Mediterranean to Latin America: The Fat Olive, a Greek trattoria, Yucca, a Latin micro-lounge, 12 Chairs, a new premium dining and The Funky Chicken, a new venture in chic fast food bringing humor and fun to retail dining. The four projects are housed together in a prominent, traditionally-styled building in Shanghai's French Concession.
思南公馆 David Laris 概念餐厅是四个不同风格的概念餐厅，灵感来源各异：从地中海到拉美风格；The Fat Olive，是一家希腊风格休闲小馆；Yucca，是一家拉美风格休闲小酒吧；12 Chairs，是一家新的高级餐厅；The Funky Chicken，是一家充满幽默风趣的新餐厅。四家餐厅都坐落在优雅传统的老法租界的一栋洋房里。

The project was a challenge uniquely suited for Thomas Dariel: maintaining the unique identity of each concept demanded a flexibility not often found in designers. Thomas Dariel refuse to adhere to one style, rather they see themselves as curators selecting an eclectic mix of elements to create environments.
对 Thomas Dariel 来说本项目的挑战在于：保持每个餐厅的特色要求灵活性，而这种灵活性并不常见于所有的设计师。Thomas Dariel 拒绝沿袭某种单调的风格，他们兼收并蓄，选择不同的元素混搭，来打造独特唯一的空间环境。

The Funky Chicken, brings chic humor to fast food roast chicken. Inspired by cartoons and wry wit, Thomas Dariel used a bold color palate and graphic architectural elements to create a fun, stimulating retail environment. Graffiti shouts "what the cluck?" on one wall while the opposite one is stenciled with cartoon speech bubbles featuring

an unexpected conversation between two chickens. The prominent red central seating counter spirals up and across the restaurant, evoking both elements of traditional Western farm architecture and flight.

Thomas Dariel 这次给我们带来了时髦的烤鸡快餐——Funky Chicken。受漫画和搞笑智慧的启发，Thomas Dariel 在作品中大胆用色，用夸张的构图和建筑元素打造出跳跃喷张的零售空间。墙上的涂鸦写着："乱叫个啥"，另一面墙上是几个卡通对话框，两只鸡之间的"胡言乱语"穿插其中。正中的大红色桌台旋转而上，焕发传统西方农庄的味道和飞翔的感觉。

This one building offers a whole day's eating: Whether it's lunch at The Funky Chicken, drinks at Yucca after dinner at The Fat Olive, or a banquet at 12 Chairs, these four concepts represent a new landmark for design and food in Shanghai.

你可以在这幢楼里吃一整天：无论是在 The Funky Chicken 里吃午餐，在 The Fat Olive 吃过晚饭后去 Yucca 小酌抑或是在 12 Chairs 里盛宴，这四家餐厅已成为上海设计与餐饮的新地标。

YUCCA
尤卡餐吧

Design Company : Dariel Studio
Designers : Thomas DARIEL
Location : Shanghai
设计公司：Dariel Studio
设计师：唐启龙
位置：上海

Yucca is located at Mansion 26F in Sinan Mansions, in a prominent, traditionally-styled building of the former Shanghai French Concession. The building is the headquarters of Yucca's creator, Australian-Greek renowned Chef David Laris, and it also houses three of his other restaurants: The Fat Olive, a Greek trattoria, 12 Chairs, a new premium dining and The Funky Chicken, a new venture in chic fast food. Perched on the third floor, Yucca is a swanky Mexican lounge featuring modern chic atmosphere and eyecatching décor with a mismatch of styles and colors.

Yucca 坐落在思南公馆26F，在上海老法租界的一幢醒目的传统风格的建筑中。这幢楼是 Yucca 的创始人、著名的希腊裔澳大利亚厨师 David Laris 的总部，也涵盖了他的另外三家餐厅：The Fat Olive, 希腊休闲小馆；12 Chairs, 新的高级餐厅及 The Funky Chicken, 一家时髦快餐厅。坐落在三楼，Yucca 是一家时髦的以现代摩登氛围和抓人眼球的设计装修为特色的墨西哥餐厅。

The word "Mexican" vis-à-vis restaurants and bars tends to conjure up a host of tired old clichés and stereotypes: cacti, sombreros, holsters, bandoliers, cracked stucco exposing faux adobe bricks or the obligatory Frida Kahlo portrait for those places attempting to "class it up". The designer - Thomas Dariel - behind Yucca did not want any of that. Far from being folkloric, Yucca pays homage to the rich visual cultures of Latin America and the Iberian Peninsula.

"墨西哥"这个词和餐厅、酒吧联系起来总会让人想到一系列固定的套路：仙人掌、宽檐帽、插袋手枪、子弹带、暴露出人造砖块的裂开的灰泥墙或者为了所谓的"上档次"而有必要挂起来的 Frida Kahlo 肖像画。设计师－Thomas Dariel－Yucca 背后的创意者不想要其中的任何一个概念。区别于一般的老套，Yucca 向拉美和伊比利亚半岛丰富的视觉文化致敬。

Fashionably Latin, Yucca's quirky yet chic and sleek interior design features a rich combination of shocking colors. The walls are splashed in vivid hues of blue and pink, the floor is a study in op art with a mosaic of tailored blue and white tiles in geometric formations laid out at haphazard angles. A photograph of blue-hued women festooned with Moorish designs extends from the entrance up the double-high wall to the top-floor lounge. A spiral staircase leads to a tiny lounge area for private parties where ones can overlook the entire space and feel the vibes coming up.

时髦拉丁风，Yucca 时髦华丽的室内设计以丰富的、有冲击力的色彩搭配为特色。墙面用生动的蓝色和粉红粉刷，地板使用了随机铺设的蓝白相间的几何图案马赛克。一幅摩尔人风格的蓝色女人照片从入口处的双高墙面一直延伸到顶层的包间。旋转楼梯引向顶楼私密空间，在这里你可以俯瞰整个酒吧，感受热烈的气氛。

The glowing glass-topped marble bar with the inscription "DESEO FUERZA AMOR LUJURIA" embodies the spirit of the space. Color combination, pattern on the floor, natural light arrangement, etc.: working together, these elements promote close interaction among the people enjoying their time in this room. Definitely for social purposes, it is as cheerful as inspiring. Intimate, Yucca is a place intended for mixing and mingling.

色彩搭配巧妙，地板图案新奇，自然光布局合理……各种要素相映成趣，使来宾在房间里更为怡然亲密。一个畅快灵动的社交场所油然而生。Yucca, 亲密无间，融会贯通。

"We imagined an exuberant place where one can go to be inspired, a place where your imagination can run free and you can be inspired by your friends," Yucca's crazy, modern Mexican feel conjures up thoughts of Salvador Dali, Diego Rivera and Frida Kahlo.

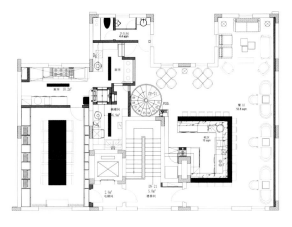

"我们希望打造一个能激发灵感的场所,在这里,想像力可以自由驰骋,朋友们也可以互相寻找灵感"。Yucca 迷人的现代墨西哥气息不禁让人想起萨尔瓦多·达利、迭戈·里维拉和弗里达·卡罗等艺术家。

Religious undertones, spanish-style custom-made mosaics, vibrant and eclectic blues tall candelabra, catholic crosses, elaborate flooring, paisley armchairs, iron gates, a sheer number of elements bringing this spirit of exuberance while being warm, cosy and harmonious.

其宗教基调、西班牙样式的定制马赛克、鲜明不拘一格的蓝色、枝状大烛台、天主教十字架、精制地板、螺旋纹手扶椅、铁门等一系列元素打造了一个多元和谐而又温暖舒适的 Yucca。

12 CHAIRS

12 Chairs, is a new premium dining concept and an intimate space where private parties can choose from menus ranging from the elegant to the extravagant. This flexibility meant designing a room to reflect and be transformed by the food. The namesake 12 chairs surround a long saddle-stitched black leather table, and burnished silver metallic panels inset with bevel-edged copper mirrors line the walls. The table settings emphasize the mood of the meal, either through the elegant place settings or over-the-top centerpieces. The effect is of being inside a crystal glass and a world of one's own choosing.

12 Chairs 是一家高级晚宴概念餐厅，可为私人派对提供私密空间。来宾可以从优雅到奢华的菜单中选择自己所爱。服务上的灵活性意味着房间的设计要反映餐品的内涵，适应不同餐品的风格。12 Chairs，顾名思义，有 12 把椅子围绕着一张依照马鞍缝制的黑色皮质桌台放置。同时，墙面饰有内嵌斜角铜镜的磨光银色金属板。优雅的餐台布置和非凡的餐台装饰无不强调着雅致的用餐气氛。来宾像置身于水晶杯中，猗伴在自己挑选的个人世界中。

THE FAT OLIVE

The Fat Olive invites the feel of a Greek home. Inspired by rustic European market towns, Thomas Dariel went about designing a casual-chic space aimed at comfort and relaxation. "We wanted the place to feel like home and so we looked at materials and details that have texture and history." Thomas Dariel achieved this feeling by carefully selecting each design element, from the flooring made of wood reclaimed from turn-of-the-century houses to the wall of European market-inspired cabinets detailed with miss-matched hardware. The predominance of earth-tone colors and copper gives to the venue a natural atmosphere while big plush bourgeois sofas in mauve add an eclectic, lived-in look.

Fat Olive 引入了希腊的家居设计风格。受欧洲乡村集市的启发，唐启龙此次追求的是舒适而放松的休闲时尚风。"我们想要创造宾至如归的感觉，所以我们选用的材质和细节都有一定的质感和历史感。"从强调世纪之交家居风格的木质地板到欧洲集市风格配以错落酒具的酒柜墙，唐启龙精心挑选每个设计元素，完美呈现了这种设计风格。土地色调的彰显和铜材的加入带来了自然的气息，而淡紫色的大长毛绒中产阶级情调沙发则散发出折衷主义慵懒的生活气息。

THE PURPLE ONION RESTAURANT

博安熙餐厅

Design Company : Dariel Studio
Designers : Thomas DARIEL
Location : Shanghai
设计公司：Dariel Studio
设计师：唐启龙
位置：上海

The Purple Onion 坐落在老法租界。周边是令人愉悦的住宅及零售区域、中式和法式建筑、国际餐厅及小店，传统的弄堂、沿街绿树成荫……今日城市构成元素的完美结合。设计师 Thomas Dariel 希望 Purple Onion 与这些元素互动，在这片颇具历史感的区域显得既传统亲切又卓尔不群。

由世界知名大厨 David Laris 提出概念的 The Purple Onion 餐厅的设计灵感由在南欧流行的颇具家庭温暖氛围的法式 Bistrot 及意大利式 Trattoria 激发。

设计体现了这些独特空间的密码而又更加深入、鲜明地重新诠释了其通常的特质。例如，像 Bistrot，是居家的感觉，你可以在这里会熟人，The Purple Onion 陈列了一系列祖先的油画肖像，目睹时光流逝 这些人物都拿着一杯葡萄酒或者鸡尾酒啜饮，微笑看着来此的客人并与大家共享欢乐休闲时光。传统家庭肖像的装饰保持了家庭氛围又带来一丝幽默与现代感。

同样地，在骄傲地悬挂于壁炉上方的墙上的巨大黑色鹿头吸引了众人的目光。这个雕刻作品再次让人联想到在欧洲家庭中普遍存在的动物头像，在这里是黑色陶瓷质地，它是一个纯粹的设计单品。

空间用三种颜色处理，黑色、紫色和白色。黑色，引领了餐厅的主要部分，结合家庭氛围的壁炉传达了私密、舒适的家庭感觉。黑色表达了潮流及优雅。在酒吧区，紫色为酒吧区特别打造的慷慨精致的色彩带来温暖、好客及迷人的氛围。白色中和并凸显了黑色紫色这两种主导色彩，垫子和光泽材料的使用也是这样。

除了酒吧和餐区，The Purple Onion 还有令人惊艳的带有林荫的庭院，隔开街道，让客人尽情享受老式建筑的外部风情。镶嵌在弄堂的中心，Purple Onion 隔离了这喧嚣城市中的疯狂，像一座宁静的岛屿，给人以家的温暖，同时又非常时髦迷人。整体设计通过体现其文化、符号和性格而赋予了 The Purple Onion 强烈的可识别性 从而将其打造成极具个性的独特空间。极具个性的美味餐点由天才厨师 Lex Hauser 主理。

KARTEL WINE LOUNGE

卡特尔酒廊

Design Company : Dariel Studio
Designers : Thomas DARIEL
Location : Shanghai
设计公司：Dariel Studio
设计师：唐启笔
位置：上海

Kartel 是一座三层楼的豪华酒吧，位于前上海法租界的核心地段。该地段是一个宜人的居住和购物区域，中欧风格的建筑，国际餐厅和小商铺，梧桐树成荫的街道和老式弄堂在这里完美融合，这就造就了今天的上海。

随着"摧毁别致"的概念，设计师唐启龙希望使 Kartel 在这片充满历史记忆的地域里，展现其优雅而又发人深省的强烈对比。

由于 Kartel 所在的大楼仍在进行之前的拆除工作，这更自然地契合了"摧毁"这一主题。伴随着破旧墙面的剥落，大楼本身的结构逐渐显现出来，更突显了应当被保留的形状、质地以及深厚的历史记忆。在欧洲，"美术"（Beaux Arts）风格向来推崇那些剥去一切掩盖其原始建筑的楼宇，因此设计师决定在此也顺势保留几处原始的风味。剥落的墙壁、未包裹装饰的柱子、建筑工人留下的字迹，都成为这些原始风味的历史元素。

为了平衡这一原始的一面，设计师特别运用优雅醒目的装饰和时尚的定制家具来营造 Kartel 充满欧亚风情的温暖氛围。酒吧一共有三层，两层室内以及一层屋顶露台，这三层不同风格的空间分别演绎了不同的情调氛围。

4 楼的布置和壁炉令人想起传统巴黎沙龙的温馨气氛，喝上一杯勃艮第红酒，与三五知己时而热烈时而亲密的聊天。

5 楼是酒吧的主要区域，透过环绕的落地窗坐拥令人惊艳的上海摩天美景，完全颠覆对于酒吧的老套印象。这里找不到传统酒吧惯用的装饰橡木桶，而会发现很别致的浴缸形状的沙发椅，配上镀金扶手，与受装饰艺术启发的桌子相符相呈。顺着墙，一大排的酒架成为整个房间的亮点，像是舞台上的杰作，让人突然回神，酒，在这里，仍然是主题。

屋顶露台上，舒适的沙发和高档的桌椅摆放适当，不管坐在哪个位子都可以拥览无与伦比的法租界 360 度全景，对于那些寻求休闲氛围的人来说这是一个绝佳的选择。

DANNY FANG

Create Love, Not More

创造爱·宁爱勿滥

DANNY FANG
丹尼·方

Dutchman Danny Fang (1975) graduated from the renowned Design Academy Eindhoven in 1998. After working as a freelance designer and developing a project for fair trade in Kenya and Zimbabwe, he worked at the Dutch Institute of Applied Scientific Research for almost two years. From 2000-2006 Fang worked at Marcel Wanders Studio where he was responsible for the product department.

In 2007 Fang moved to Hong Kong where he set up Fang Studio Ltd. In 2008 he was elected by Perspective Magazine as one of the "40 under 40" talented and young individuals who lead Asia's design industry.

荷兰设计师丹尼·方（出生于1975年）于1998年毕业于著名的艾恩德霍芬设计学院。他起初是一名自由设计师，在设计了肯尼亚和津巴布韦的一个公平贸易项目后，他在荷兰应用科学研究协会工作了近两年。2002年至2006年，丹尼·方在马塞尔万德斯工作室工作，主管产品部门。

丹尼·方于2007年移居香港，并在香港成立了方工作室。2008年，他被《透视》选为亚洲设计界"40位40岁以下"的青年才俊。

CHANGE EVERYDAY
我们需要每天改变

I believe that design needs to contribute to the world in an inspiring, comfortable and attractive way. As designers we have several responsibilities:

- design a good product

- find the most efficient manufacturing process

- be aware of the social and environmental impact of what is made.

Everyday our needs change because our environment evolves. To give focus and relevance to our projects I study the way we live, the way we work and the way we interact with each other. Through this research I come up with well-founded ideas and concepts for new products.

I design my products in such a way that we make the optimum use of our production methods. Every process and material has its own characteristics and qualities and we need to utilize these possibilities in our design language. Designing accordingly is the best way to find the balance between the production, costs, the use of available resources and the final design.

Not only we use the industry to manufacture our products and ideas, but designers also have the responsibility to guide in its development; we need to push craftsmen to develop their crafts and manufacturers to push their technology further to make more intelligent products.

In my studio we work mainly with Asian manufacturers and try to develop their specific strengths and thus stimulate development of the labor force. On the long term this will help to evolve and mature the design and manufacturing industry. With mass production everything we do gets multiplied. Every unsafe or polluting production method gets multiplied. On the other hand, every added value we introduce, every resource we save and every improvement that we make will multiply as well. In my eyes this awareness is the best way for designers to contribute to a more sustainable world. When we focus on creating out of love, I believe profit will come. That is why my motto is "Create love, Not more."

我认为设计需要以鼓舞人心，舒适和吸引人的方式向世界作出贡献。作为设计师，我们肩负着几大责任：

- 设计一个好的产品

- 找到最有效的制造过程

- 充分认识到制造出来的产品可能产生的社会和环境影响

每天，我们周围的环境都在不断发生演变，我们也因此要做出相应变化。为了找到项目的重心和关联性，我对我们的居住方式、工作方式和交流方式进行了仔细研究。通过研究，我为新产品找到了确切的概念和贴合的思路。

我设计产品时，会力图使我们的生产方式得到最佳利用。每一道流程、每一种材质都有其自身固有的特点和特质，我们需要在我们的设计语言中充分利用这些可能性。据此得出设计方案，是平衡生产、成本、现有的资源利用和最终设计方案等各项要素的最好方式。

设计师不仅是要利用工业来生产我们的产品，让我们的想法变为现实，同时，还有责任引导产品的发展：我们需要鞭策工匠们不断完善自己的技艺，鼓励制造商们不断改良自己的技术，使产品更为智能化。

我的工作室主要是和亚洲制造商合作，并尝试开发其特定的优势，从而刺激劳动力的发展。长期看来，这将有助于发展成熟的设计和制造行业。随着大规模生产的开展，我们得到的一切都会翻倍。每一种不安全的、会产生污染的生产方式都会翻倍。另一方面，我们引入的每一个附加值，我们节约的每一项资源和我们所作出的每一项改善也将成倍增加。在我眼里，能有这种意识，对于设计师来说，是为建设一个更环保的地球做贡献的最好方式。当我们用心来创作时，我相信，利润就会随之而来。这就是为什么我的座右铭是"创造爱，只要爱"。

INTERVIEW: DANNY FANG

丹尼·方访谈

5Cats: You were born in the Netherlands and studied in Eindhoven. You came to Hong Kong in 2007. In a few years' time, how did you bring your design ideas to Hong Kong and make Hong Kong accept it?
5只猫：您出生于荷兰，在埃因霍温学习，07年来到香港，短短几年的时间，您是怎么把您的设计思想带到香港，并让香港接受的呢？

Danny Fang: When you talk about design ideas, do you talk about style or philosophy?
丹尼·方：这里的设计理念，指的是设计风格还是设计思想？

Most people talk about style as an outer appearance, so normally I try to avoid that discussion. However, if you talk about my style I would say it is clear, friendly and pragmatic. I always try to find ways to display the available qualities of an object, its material and its way of manufacturing in combination with functionality and esthetics. I have discovered that if you stick to these rules, people from all over the world can appreciate it.
丹尼·方：大多数人谈到设计风格的时候，一般指外观，所以通常情况下我都避免讨论这个话题。但是，如果您指的是设计风格，我会说干净、友情和实用。我总会尝试通过各种方法，结合功能和美学，展示某项物体可以利用的品质、材质和制造方法。我发现如果你坚持上述原则，就会得到全世界人民的欣赏。

More difficult to get explained is my design/business philosophy:'create love, not more'. What it comes down to is that I feel that every project I do should be something I believe is worth sharing with the rest of the world and the process to get there has to be sustainable. This means that there is a need for more critical thinking on every level and a more long-term strategy from my clients. I have been lucky that when I first came here, there were people who were willing to give me a chance to prove myself and luckily enough I am finding more and more clients who support the philosophy. I always look for people with similar mind set, so we can grow together.

丹尼·方：要解释我的设计/商业思想，可能比较难："创造爱，别无其他"。这是因为，我认为我做的每个项目都应该是我认为值得和世界上其他人分享的东西，且其间的过程是持久的。这意味着我的客户在各个层面都进行更为批判性的思考并拥有更为长期的战略。我很幸运，当我第一次来到这里时，就有人愿意给我机会证明自己。更为幸运的是，我找到越来越多支持我这种设计理念的客户。我总是寻找与我志同道合的人，这样我们可以共同成长。

5Cats: I heard you once designed high-end chairs for a local company. One of them got inspiration from the classical chair in Ming Dynasty. Could you give us more details?

5只猫：听说您曾经为内地的一家公司设计过高档椅子，并且其中一款是从古典的明朝椅子中吸取的灵感，能聊聊其中的过程么？

Danny Fang: Two years ago I met my friend Elvin Huang. He owns a factory that makes plastic chairs and he is specialized in polycarbonate chairs that are made by gas assisted injection molding. He asked me if I could design a chair and take it in production with him. For me this was a great opportunity, because it is seldom you get the chance to design something in this high-end technology and to develop such big tools. In traditional plastic chairs the technology restricts the design to the same material thickness all over the product. But the beauty to the gas assisted technology is that you can 'cheat' the rules by blowing an air bubble inside so you can create a visual thicker part. To me the Geppetto chair is a tribute to the Ming chair; inspired by the language of the chairs I created a structure that looks solid but with the soft and friendly transitions of a Ming chair. By using the liquidity of the plastic when it is injected I have the seat, back and armrests flowing out of that structure, something that would be almost impossible to do in wood. The funny comparison you can make is that both the Ming chair and mine do not use any glue or screws. The difference is that mine can be made in only 5 minutes.

丹尼·方：两年前，我结识了我现在的朋友Elvin Huang。他开了一家生产塑料椅的工厂，利用气辅注塑技术，专做聚碳酸脂座椅。他问我是否能设计一种可以和他合作生产的椅子。对我来说，这个机会非常好，因为很少能有机会用这种高端技术设计作品并推动如此大型的工具的发展。传统的塑料椅工艺要求设计的产品各处的材质厚度相同。但是气辅技术的魅力就在于你可以在内部吹入气泡而"骗过"规律，由此产生视觉上的薄厚之分。对我来说，Geppetto椅是向明代椅子的一种致敬：受椅子语言的启发，我创造了一种看上去坚固但实为对明椅的柔软而友好的转变的结构。利用塑料的流动性，在注塑过程中，从结构中流出了座椅、靠背和扶手，而这些用木制材料几乎是无法实现的。有趣的是，你会发明椅和我的作品都没有用胶水和螺丝。不同的是我的作品只需要5分钟。

5Cats: You once said, "Chinese enterprises are flourishing, while lack of proper guidance. As a industrial designer, I can bring them great changes." On which aspects do you think can your designs boost the enterprises? How to make an enterprise approve a designer, and approve that you have values for them?

5只猫：您曾经说过"中国的企业蓬勃发展，但缺乏正确的引导，身为工业设计师，我能给他们很大的改观。"您认为设计能从哪些方面带给企业助力？如何让一个企业认可一位设计师？认可你们能带给他价值？

Danny Fang: You have to admit that is quite impressive when you come to Hong Kong or Shanghai and you see the thousands of filled containers being handled and distributed all over the world every day. But up to now China has mainly been flourishing as an executing manufacturing power. With the growing labor costs and fierce competition I see a growing need for the industry to become original design manufacturers and brands. So there is need for people who can help these companies finding their path into the world of creative thinking and design. I can help them to change their focus and help them design product ranges that gets them in the right direction.

丹尼·方：必须承认的是，当你来到香港或上海的时候，你会看到成千上万满载货物的集装箱每天被运发到世界各地，并因此留下深刻印象。但至今为止，中国还主要作为一个执行生产大国来发展。随着劳动力成本的增长和竞争的加剧，我看到产业对于原创设计制造商和品牌愈发强烈的需求。因此，需要有人能够帮助这些企业寻求跻身这个充满创新思维和设计的世界的途径。我可以帮助他们转移重心，帮助他们重新设计产品的范围，找到合适的方向。

For companies to approve a designer has everything to do with expectations. When I was at the Shunde design fair last year, there were a couple of factory owners who came up to me and asked: "how much for a design and how much if I buy three?". When I asked what they were looking for they had no clue.
So in order to judge if a designer is the right one for you, you need to be able to express what you expect from them. The same goes for me when I work with a new client. I always ask what they expect from me and where they want to be in say five to ten years. This is important because that way you can make the right strategic decisions. That is a big part of my added value; next to my creativity and technological knowledge, I have a good understanding of the business side of things.

丹尼·方：对于要验证设计师是否完全符合其期望值的公司来说，去年，当我还在Shunde（尚德，音译）设计公司的时候，有一些工厂老板找到我，然后问我："一个设计多少钱，如果我买三个多少钱？"当我问他们想找什么样的设计的时候，他们完全没有想法。所以，要判断设计师是否是你想找的，你要能说出你希望他们给你什么。我寻找新客户的时候也是一样。我通常会问他们期望从我这里得到什么，比方说，五到十年之后，他们想做到什么程度。这一点非常重要，因为这样你可以做出正确的战略决策。这也是我除了创新和技术之外的附加值中的很大一部分——我非常了解事物商业的一面。

5Cats: Some designers seem to be too "highbrow". Their works are very artistic but difficult to be put into business production. While your designs are good examples balancing business and arts. Could you tell us some stories in this?

5只猫：有的设计师似乎"曲高和寡"，他们的作品有着艺术价值却很难投入商业生产，而您的设计却在商业化和艺术品之间找到了一个平衡点，您能说说其中的故事么？

Danny Fang: Art is a lot more autonomic than design: art is very research driven, critical and should not be restricted by a commercial outcome. There are many designers who prefer to work in this way. But I personally love the puzzle of bringing together disciplines like philosophy, engineering, marketing and logistics. I flourish when I am in a factory with the craftsmen, finding their specific skills, implementing innovation. Having an artistic drive in combination with commercial targets help me focus and make intelligent decisions. For example, I would never frustrate the production line or make a product 50% more expensive just because of a minor design detail. If this implies the product doesn't look nice anymore, I get forced to be more creative and come up with different solutions.

丹尼·方：与设计相比，艺术更具自发性。艺术很大程度上受研究驱使，更为批判，不应受到商业结果的限制。很多设计师也更乐于此道。但从我个人来讲，我更喜欢谜一样地将多学科比如哲学、工程学、市场影响和物流结合在一起。当我在工厂和技师在一起的时候，会发现他们独特的技能，并将创新付诸实践，这让我感到无比兴奋。艺术的冲动加上商业的驱使帮我找到重点并做出明智决策。例如，我不会仅仅因为一个很小的设计细节而放弃生产线或者让产品价格高出50%。如果这样有可能影响产品的美观，我会努力地用更加创新的方法找到不同的解决方案。

5Cats: The Netherlands is a small country, but it is the design kingdom. The competition is intense. So Dutch design companies all find their own core and remove the non-core stuff, to produce the most professional designs. Many Chinese designs feel so full, with so many elements, thus losing core. You once said such words. Does that mean the simple design is the best design?

5只猫："荷兰国家小，却是设计王国，竞争更加激烈，因此荷兰的设计公司都会找准自己的核心，将非核心的东西去掉，做最专业化设计。中国很多设计会让人觉得很满，元素多了，核心就不突出了。"这是您曾说过的话，是否意味着，简洁的设计才是好设计呢？

Danny Fang: Oof, did you ever walk over a Chinese trade fair? There is so much, and it looks like everyone makes the same products. So you have no idea where to start. And when you look at many of designs they make they are so full and un understandable that I can't see if people are trying to design a chair or a sport shoe!

丹尼·方：是啊！你去过中国的展会么？规模不小，但是似乎大家的产品都一个样，所以你也不知道从哪看起。当你看了很多很满又让人费解的设计之后，你甚至不知道他们是要设计椅子还是运动鞋了！

Personaly I always compare design with writing poetry. Poetry can be very simple as well as very rich. A product on itself is a visual sentence. You have a primary thing you want to say, a subject. And every detail on the product, just like every word in a poem has a meaning, a reason. Everything that distracts from that, you should leave out. Your message becomes unclear. This is the difficulty of making a good design.

丹尼·方：从我个人来讲，我通常将设计比作作诗。作诗可以非常简单同时又非常丰富。产品本身就是一个视觉上的句子。你有你想说的核心事件——主题。而产品的每个细节，就像诗里的每个单词，都有意义，都有原因。任何偏离主题的东西都应该摒弃，会浑浊你的信息。这就是好的设计的难点。

Design companies have a similar problem. You can see the collection of products as a story. The products tell about the strength and the values of a company. If you want to become a brand, you need to eliminate the products that make that message unclear.

丹尼·方：设计公司也面临同样的问题。你能看到产品的系列就像一个故事，产品诉说的是一家公司的实力和价值。如果你想创造一个品牌，你就需要剔除可能浑浊信息的产品。

So it is not said that simple design is the best design, but it does need to be consistent to be understandable.

丹尼·方：不是说简单的设计就是最好的设计，但是必须要让人理解。

5Cats: HK, especially the Mainland has great cultural differences compared with the Netherlands. Could you use your own experiences as examples to tell us, what is the biggest difficulty to overcome when a designer comes to a culturally totally different country for work?

5只猫：中国香港，特别是内地，和荷兰的文化差异还是很大的，能否以您的切身经验谈一谈，设计师来到一个完全不同文化的国家来做设计，需要克服的最大困难是什么？

Danny Fang: Despite what everybody thinks, we are not that different; both in Holland and China the boys are running after the girls and the girls are running after the boys and we all try to get ahead. So the biggest difficulty to overcome is your own prejudice. In Holland we are taught how things should be organized from a social, economic and technical aspect. What many foreigners forget is that we reached that structure after 150 years making mistakes, learning, emancipation battles, worker strikes etc. China has a totally different history and is still an emerging country. To work and live here you need to let go of your own tradition of thinking and expectations and become sensitive to the practical problems China has to deal with.

丹尼·方：尽管所有人都这么认为，但是我们不是那么不同，荷兰和中国的男孩都在追求女孩，而女孩也在追求男孩，而我们都想处于领先地位。要克服的最大困难就是你自身的偏见。在荷兰，我们受的教育是从社会、经济和技术的层面组织事务。很多外国人忘记了我们是在经历了150年的犯错、学习、解放战争和工人罢工等等才认到这个结构的。中国的历史完全不同，中国仍然是个崛起中的国家。在这里生活和工作，你需要放开自身的思维和预期传统，对中国必须要解决的实际问题变得敏感。

5Cats：Art design can achieve business benefits. We know some famous cases of Apple. So does art design purely for the sake of business have any influence on the development of art itself?

5只猫：艺术设计可以成就商业利益，我们知道苹果公司就是很出名的例子，那么纯粹为了商业的艺术设计对艺术本身的发展有什么影响吗？

Danny·Fang: Apple was the first to use human centered designers to design electronic products in stead of design engineers. The products were designed out of a human interest. This was simple, but eye opening to the rest of the world. Apple had no other intention other than to have their design create support for doing business. But in this they proved to the world what design can do for your brand and business. I hope that companies start to understand that, stop copying the looks and start copying the strategy.

丹尼·方：苹果公司是第一家采用以人为本的设计师而非设计工程师来设计电子产品的公司。产品的设计完全出于对人本身的考虑。这很简单，但是却打开了全世界的眼界。苹果公司的唯一目的就是用设计为商业创造保障。但此时，他们向全世界证明，设计可以给你的品牌和商业做些什么。我希望公司都开始懂得这一点，停止对外观的抄袭，而开始对策略的抄袭。

But i am not sure if it has influence of the art itself. The art of design is a complicated puzzle. I feel work like Apple's should inspire designers to do their work better. But still many mediocre designers just copy the style. So as much as i like it to be, i don't think it has to much influence.

丹尼·方：但是我不确定这是否对艺术本身有所影响。设计的艺术就像一个复杂的谜。我觉得类似苹果公司一样的设计工作会给设计师带来启发，让他们在工作中表现得更为出色。但是仍有很多平庸的设计师只会抄袭别人的风格。所以尽管我很喜欢，但是我不认为会有很大影响力。

5Cats： Which direction do you think the future design is going?
5 只猫： 您觉得未来的设计是朝什么方向发展的？

Danny Fang： The last 15 years we have had, as I like to call it, a Conceptual Revolution. Design schools have been following the idea that everything has to become conceptual and designers hardly know how to make a technical drawing anymore when they leave school. It is a big difficulty for the industry to have to train these youngsters the basic skills we need for the industry to work. But the good thing is that that generation is more aware of the bigger idea behind what we are doing as a society and where we should be going. Design can take an active role in the transformation from a heavy polluting consumer society, to something more sustainable. Products will need to become less trendy and fashionable, but the service around them will be. For instance, look at smart phones: It is not the phones that create the industry; it is all the applications and services we can get through them. In the future I believe designers will help create more service-orientated industries, creating added value without necessarily using more means.

丹尼·方：过去我们经历的这十五个年头，我愿意称之为概念革命。设计院校一直崇尚的理念是，每一样事物都应该成为概念性的，设计师在离开学校的时候甚至不再知道如何画出技术图纸。业内的一大难题是必须训练这些年轻人在业内工作所需的基本技能。但优点是，这一代人对于我们作为一个团体正在从事的事情以及我们前进的方向背后的更大的概念更有想法。设计可以在从污染严重的消费者社会到更为可持续型社会的转型过程中发挥积极的作用。产品将需要变得少一些新潮和时尚，但是围绕它们的服务将需要更为时尚。比如说，智能手机：不是手机本身创造了这个产业，而是我们可以在手机上应用的所有软件和服务创造了这个产业。我相信未来的设计师会创造更为服务化的产业，而这不一定要寻找更多的方式就可以创造附加值。

5Cats： Do you believe design has a trend? Is trend made by new material that is constantly updated with the

development of science and technology?
5只猫：您觉得设计有潮流吗？是否说潮流是随着科技的发展，不断产生新的材料导致的？

Danny Fang: Design is a good measure where we are with our level of development of new materials and technology. But also were/who we are as a society. So looking at design, you can learn a lot about how developed and sophisticated we are. What do we do with new technology? Do we design new weapons, or do we design to make our lives better? And how effective are we in achieving that? So every time science comes up with something new, designers get challenged to give it an intelligent application. That is the fun of what we do.
丹尼·方：设计是新材料和技术发展水平的很好的标杆。同时也是我们整个社会身份和方向的标杆。所以当我们看设计的时候，我们就能知道我们有多么的先进和发达。我们要拿新技术怎么办？我们要设计新的武器，还是我们要把生活变得更加美好？我们达到这个目标有多有效率？所以每当科学有了新发现时，设计师都要迎接挑战，把它变为一项智能的应用，这也是我们工作的乐趣所在。

5Cats: Do you have favorite designers in HK? Compared with Europe, what is the deepest impression that HK local designs give you?
5只猫：在香港您有没有喜欢的设计师？和欧洲的设计比起来，香港本土设计给您留下最深印象的是什么？

Danny Fang: In Hong Kong there are many who do their own little thing and I love them all for that. I respect all those who take our profession seriously and take ownership for what they do. But the deepest impression I got is about the entrepreneurial attitude here in Hong Kong. Everybody has his or her own little production or side business going on. That is very inspiring for someone who has been taught that the cobbler should stick to his last.
丹尼·方：香港有很多做小设计的设计师，我对此非常欣赏。我尊敬所有对设计这个职业持严肃态度并对其从事的工作负责任的设计师，而给我留下最深印象的则是香港设计师的企业精神。大家都有自己的小产业或正在从事的副业，这对于被教育要"安分守己"的人来说非常具有启发性。

ALUMINUM FRAME, POLYETHYLENE WICKER

俄罗斯套娃椅

Project Name: aluminum frame, Polyethylene Wicker
Design: Danny Fang, Fang Studio
Client: Kian
Date: 2007
项目名称：俄罗斯套娃椅
设计者：丹尼•方·方设计工作室
客户：基安 SDN BDH（马来西亚）
时间：2007

The 'Matryoshka Chair' is a light weight outdoor furniture set consisting of two chairs and two side tables/stools. Inspired on the traditional Matryoska dolls the Papa chair can open up, and inside the Mama chair appears. In the Mama chair there are two children hidden. Nice detail is that one can still sit on the Papa Chair when the whole set is stacked.

"俄罗斯套娃椅"是一套轻便的室外家具，包括两把座椅，两张桌子/两张凳子。该作品的灵感来源于传统的俄罗斯套娃，将爸爸椅打开，就会看见里面的妈妈椅。妈妈椅里面藏着两把娃娃椅。作品中，一个很好的细节设计就是，当把整套椅子都收起来后，你还可以坐在爸爸椅上。

GRADIENT COLLECTION
错落有致

Project Name : Gradient Collection
Design : Danny Fang, Fang Studio
Client : Kian
Date : 2008-2010
Material : Bottom frame, Galvanized steel with outdoor powder coating. Seat, aluminum with polyethylene fiber
项目名称：错落有致
设计者：丹尼•方．方设计工作室
客户：基安 SDN BDH（马来西亚）
时间：2008-2010
材料：底框，带户外粉末涂层的镀锌钢．底座，铝聚乙烯纤维．

Winner of the bronze award at Design For Asia Awards, 2010, Household Appliance - Product Design
Winner of a special merit award at perspective awards 2010, Category product design.
荣获 2010 年度亚洲设计奖家电产品设计类铜奖
荣获 2010 年度透视设计奖分类产品设计特殊贡献奖

The Gradient Collection is a set of all-weather wicker chairs that is a contemporary interpretation of traditional wickerwork. The collection is a showcase of the unique quality of the age-old craft of wicker weaving through a visually stunning hand woven pattern.
错落有致产品系列是对传统柳条工艺的现代诠释。该系列通过视觉上惊为天人的手工编织图案来展现古老的柳条编织手艺的独特质地。

The Gradient Collection transforms the craft of weaving by discovering new possibilities in hand woven designs. Traditionally, wicker craftsmen are seen as mere manufacturers of a functional object, but this collection places the spotlight on the craftsmen by emphasizing the beauty of a weave of exceptional quality.
错落有致产品系列发掘出手工编织设计的最新可能，并将其融合成新的编织工艺。传统上，柳条编织艺人只是做一种功能物件，但是这个系列的产品更强调超高品质编织的美感，将手艺人放到了聚光灯下。

CHECK MATE!

将军！

Product Name : Check Mate!
Design : Danny Fang, Fang Studio
Client : HIVE
Date : 2011
Material : resin / paper / gold leave
项目名称：将军！
设计者：丹尼•方．方设计工作室
客户：HIVE
时间：2011
材料：树脂、纸、金叶

In the game of chess, the position of all pieces is crucial to win or to loose. We want to position our lights just as careful and strategic to get the best ambiance in our environment. The Check Mate collections consist of a complete set of chess pieces that all have their own character and size. One can choose from the modest Pawn all the way up to the luxurious King and Queen to fit the position you need lit. With their warm candle like light that is being reflected of the gold leafed inside together with the tactile paper pulp skin the Check Mate collection brings warmth to any environment.

象棋比赛中，每个棋子的位置都对胜负起着关键作用。我们想从战略上认真布局我们的灯光，为环境创造最佳氛围。将军系列包括一整套各具特色和型号的象棋棋子。从最小的卒到位高权重的将，你可以任意选择，任意摆放。温暖的烛光，内藏金色的叶子，伴以触感灵动的纸浆表面，将军系列为环境带来了一丝暖意。

King

Rook

Bishop

Knight

Queen

Pawn

FANDANGO
方丹戈

Product Name : Fandango
Design : Danny Fang, Fang Studio
Client : HIVE
Date : 2011
Material : cotton/metal
项目名称：方丹戈
设计者：丹尼•方·方设计工作室
客户：HIVE
时间：2011
材料：棉、金属

I found a photo of a flamenco girl throwing her dress all around with great power and expression. This photo inspired me to make a light with similar drama and expression to spice up our modernistic world.
我曾经看到过一幅照片，照片上跳佛朗明哥的女孩尽情地甩动着裙子。这幅照片给了我灵感，我想用同样的手法和表达方式，用一点光点亮我们这个现代的世界。

To achieve this theatre and energy we had to develop new materials. Through an intricate process of working on the materials, we created petals made of fabric that we could mould in any shape, diffused the light and defied gravity.
为了达到同样的戏剧和能量效果，我们就要寻找全新的材料。在材料上大费周折后，我们最终选择轻盈弥散、形如流水的布质花瓣。

This light wants to dance in freedom, so give her space to work her gipsy magic.
这种灯光让人想自由地舞蹈，所以给她空间，发挥她的吉普赛魔力。

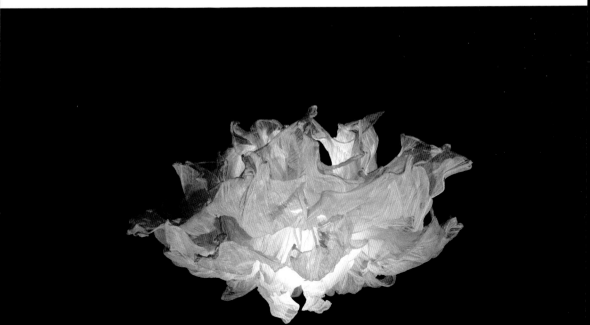

GEPPETTO CHAIR
盖比特椅

Product Name : Geppetto Chair
Design : Danny Fang, Fang Studio
Client : More Than Legs
Date : 2009
Material : polycarbonate
项目名称：盖比特椅
设计者：丹尼•方·方设计工作室
客户："居高临下"
时间：2009
材料：聚碳酸酯

About More than legs
More Than Legs is dedicated to creating beautiful and affordable high-quality plastic furniture. It is a collaboration between Dutch designer Danny Fang and Chinese manufacturer Elvin Huang, two people who are passionate about furniture. They aspire to bring good design out of China, with outstanding quality, to compete with the world's best brands.
"居高临下"简介
"居高临下"致力于创造美观价廉的高品质塑料家具。它源于荷兰设计师丹尼•方和中国制造商埃文•黄之间的亲密合作，两位创始人都对家具十分热衷。他们想要将好的、高质量的家具推出中国国门，与国际一线品牌同场竞技。

Winner of a special merit award at perspective awards 2010, Category product design.
Winner of a special Merit award at Design For Asia Awards, 2010, Household Appliance - Product

荣获 2010 年透视设计奖分类产品设计特殊贡献奖
荣获 2010 年亚洲设计奖家电产品类特殊贡献奖

Geppetto is a polycarbonate chair inspired by the ingenuity of China's Ming dynasty furniture. The chair perfectly fuses the elegant form of a classic Ming chair and the pragmatic sensibilities of contemporary design. The result is a lightweight, well stackable and weather resistant armchair that pays homage to traditional Chinese aesthetics.
盖比特聚碳酸酯座椅的设计灵感来源于明朝家具的独创性。这种椅子完美地融合了传统明椅的优美造型和当代设计的实用敏感。最终的产品轻便、易存放，受天气影响小，是对传统中国美学的参拜。

The classic language of Ming dynasty furniture is well expressed through the technology of gas-assisted injection molding, which was used to create the Geppetto.
明代家具的经典语言通过应用于盖比特椅的气注塑形工艺得到了很好的诠释。

Ming furniture is made from a single type of material without the need for any adhesives or nails. The Geppetto, being made in one piece of polycarbonate, draws an ironic parallel. The modern technology also allows the perfect curves of a Ming chair to be rendered in a split second, such that an elitist aesthetic style that used to take many days to create by hand can now be mass-produced and enjoyed by all.
该家具都由单片材料制成，无需粘胶和钉子。盖比特椅由一

整块聚碳酸酯,画出两条有讽刺效果的平行线。现代科技能够在眨眼间塑造出明椅的完美曲线,杰出的做工和艺术风格原来需要数天才能手工完成,现在却实现了大批量生产,进入了寻常百姓家。

The Geppetto is lightweight yet unbreakable and weather resistant – something that antique Ming chairs never achieved. It is also one of the few chairs with an armrest that stacks snugly. The result is a contemporary design that is a tribute to and a comment on classic Chinese aesthetics, bringing the ancient Ming designs into the present day, and breathing new energy into the traditionally wooden objects, much like Pinocchio coming to life.
盖比特椅十分轻巧,但却十分坚固,能够适应各种气候——而这是明椅望尘莫及的。盖比特椅也是少数的扶手罗列紧密的椅子之一。最终呈现的作品是一种对传统中国美学的顶礼膜拜和解说,它将明代设计带到了今天,让传统的木质结构呼吸最新的能量,就像匹诺曹复活一样。

The Geppetto is the first high-end original design for gas assisted molded plastic chairs to come out of China that is suitable for both Western and Asian markets. It's stackability, lightweight and durability makes it suitable for commercial projects as well as the private home.
盖比特椅开了气注塑形塑料椅的高端原创设计先河。它轻巧耐用,好罗列,商用家兼可。

The Geppetto maximizes the potential of traditional Chinese aesthetics while catering to the real needs of contemporary urban lifestyles. It is a contemporary comment on a traditional Chinese product.
盖比特椅将传统中国美学的潜能发挥到极致,又符合当代都市生活的实际需求。是对中国传统家居的当代诠释。

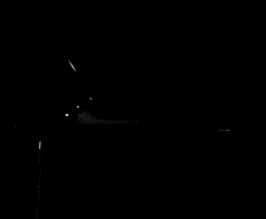

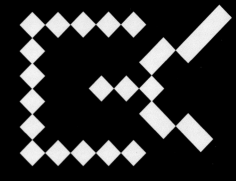

JURGEN BEY

Designed to tell
the story of language exploration

设计是诉说探险故事的语言

STUDIO MAKKINK & BEY

Makkink & Bey 工作室

RIANNE MAKKINK+JURGEN BEY

莱亚尼·马金克 + 尤尔根·贝

STUDIO MAKKINK & BEY
MAKKINK & BEY 工作室获奖情况

2009	Nomination for the Rotterdam Design Prize, for Prooff Lab
2009	普鲁夫实验室获鹿特丹设计奖提名
2008	Awarded with Woonbeurspin, for collected works
2008	集体作品获 2008 "旋转的生活" 展览奖
2007	Nomination for the Rotterdam Design Prize, for best studio
2007	获鹿特丹设计奖最佳工作室提名
2005	Prins Bernard Cultuurfonds Award, for collected works
2005	集体作品获贝恩哈德亲王文化奖
2005	Harrie Tillie Award, Stedelijk museum, Roermond, for collected works
2005	集体作品获鲁尔蒙德市博物馆哈里蒂利奖
2003	Interior award, Lensvelt/de Architect, for meeting room Interpolis 2003
2003	"城市之间" 会议室获 2003 年室内设计奖
2003	Nomination the Rotterdam Design Prize, for LinnenkasThuis
2003	衣柜房子获鹿特丹设计奖提名

DESIGN IS THE LANGUAGE OF ADVENTURE

设计是诉说探险故事的语言

Ever since the beginning of mankind, we have been trying to understand the world around us. Because of our well-developed brain, we are no longer content with just taking care of the essentials: eating and reproducing. To legitimise the way we act, we are searching for answers. These answers become stories, many beautiful stories which tell us how to act and how to look and what to see. At the time Plato lived, the swing movement of a pendulum was illustrated as the struggle of a fish. The pendulum swinging, resisting its state, to eventually get tired and give up his fight, swinging less and less. One could get enraged by all the smarty pants who would not settle for this story and keep asking questions.

自人类伊始，我们就试着理解我们周围的世界。我们发达的大脑不再满足于人类最基本的需求：填饱肚子和繁衍后代。为了让我们的行为方式合理化，我们不断地寻找答案。这些答案变成了故事，而很多美好的故事都告诉我们如何做、如何看、看什么。柏拉图的时代，钟摆的摆动被解释为鱼的挣扎。钟摆摆动时，钟摆会抵制这种状态，最后变得疲惫，放弃挣扎，并摆得越来越慢。而所有不满足于这个故事、喜欢刨根问底的聪明人都会激怒任何一个相信这种说法的人。

Photographer Kjell Sandved found the images of the man-made world in nature, on the wings of butterflies, and made them visible by filters and photography. In this project he focused on the alphabet, but with this you know that you can find every event and everybody in past and future, in a rough-resolution 2-dimensional image. The good spectator can see the world at random. The world around us can throw us into ecstasies, and it is art to see that.

摄影师谢尔·桑德菲通过滤镜和照相技术将蝴蝶翅膀上的人为世界展现在世人面前。在他的一系列作品中，集中出现了人类使用的字母。你会看到，模糊的二维图片上写着每个人每件事的前世今生。会观察的人就会看清这个混沌的世界，我们周围的世界将我们抛向迷幻，而看清它们就是一门艺术。

We feel like explorers, travelling around the world either driven by curiosity, or sent on a mission, constantly investigating, inquiring and associating. To come back with stories, stories that are told through design for design is our language.

我们就好像是受好奇心驱使或肩负使命的探险家，周游世界，不停地发现、询问和关联。我们带回故事，并通过设计诉说这些故事，因为设计就是我们的语言。

Designer Jurgen Bey and architect Rianne Makkink have operated Studio Makkink & Bey together in Rotterdam since 2002. The studio's many projects are diverse, and include public space projects, product design, architecture, exhibition design and applied arts. Aspects such as analysing the content, the relationship between purpose and shape, design as a bearer of a story, and the relationship between spectators and bearers, designer and commissioner, can originate from exhibition design and architecture, but are also used in projects of other disciplines. Urban planning, architecture, landscape architecture are inextricably bound to product design. The light bulb has had an influence on architecture, the way a house is built influences its interior, a skyscraper could have never existed without the invention of the elevator.

自2002年起，我们就开始在鹿特丹经营 Makkink & Bey（马金克和贝）工作室。工作室接收的项目风格多样，包括公共项目、产品设计、建筑、展会设计和应用艺术等。内容分析、用途和形状关系、故事性设计、观众与载体间关系、设计师与委托人等一系列问题都可以归结为展览设计与建筑，同时也可以应用于其他学科项目。城市规划、建筑、景观建筑与产品设计是密不可分的。灯泡会对建筑产生影响，房屋的建筑方式也会对内饰产生影响，如果没有发明电梯，也就不会有摩天大楼。

INTERVIEW: JURGEN BEY

尤尔根·贝访谈

5只猫：提到荷兰，人们通常会想起风车、郁金香、木屐。荷兰的特殊地理位置和独特的荷兰文化也带来了独具特色的荷兰设计。您曾提过"荷兰式的想法是干净的"，在您的设计作品中包括纽约Droog店面设计，奇思妙想的"耳椅"，这些各具特色的作品中，荷兰文化在其中起到了什么作用？

尤尔根·贝：我的父母出生在荷兰，我也由此成为了荷兰人。我觉得有点难说——你是什么背景，而这种背景对你又会造成什么影响。比如说一位作家，或者历史上的一位画家可能会有这种联系，会让你想起来这就是荷兰的风格，但是我站在我自己的文化当中，无法看到文化是怎么影响我的。我站在文化当中，很难对文化有终身性的观点，看出文化是怎么让我这么做的。比如，我之所以为我，因为你站在文化之中，所以无法总揽文化。

5只猫：您在意大利、西班牙都学习过。西方的文化、艺术、经济环境各方面都是非常联系紧密的，可以说这是一个设计师的大西方时代。而中国也有过文化的大中国时代，中国的设计师也逐渐正在显现自己对本土文化的审美倾向，这种东西方文化的碰撞有时会产生意想不到的效果，那么您在设计构思时，有尝试在设计中融合不同文化以产生火花吗？

尤尔根·贝：这个项目，我们是与上海的设计师合作的，我们用中国的元素来做与清洁有关的产品。我第一次到印度的时候，看到那么多的人在做清洁工作。在荷兰，虽然说清洁工作不是很低级或者最低级的工作，但是任何人都可以胜任。所以，人们一想到清洁工作，就只想到效率，用廉价的东西，就做就行了。我们在中国做的项目是，在靠近厦门的地方有一家做非常精致花纹的陶瓷厂，我们通过他们来做清洁用品，比如扫帚，抹布等。清洁用具是用陶瓷做的，非常精致，而且抹布采用的是带刺绣的丝绸作为材料，用他们作为清洁工具需要有能力的人，用这样的工具也能培养有能力的人。材料非常的奢华，这系列的产品都非常棒，非常精致，精致到如果你使用他们，你会用无比优雅、无比上乘的方式来操作，所以只有有能力的人才能胜任。利用中国的工艺，加上这种设计，体现出环境的优雅，然后做出了这些具体的产品。这就是我利用不同文化的一个例子。

5只猫：过去流行包豪斯或者洛可可，现在流行低碳，虽然潮流总在变化但是艺术家总能用自己敏锐的触觉来引导流行，您认为未来的设计思想将是什么？

尤尔根·贝：噢，这个问题有点难，我觉得就我而言，我在设计中有大量的工作，并且会用到大量的人手，我觉得这是一种持久的方式，让越来越多的人来工作，尽可能多地让更多人参与进来。我们采用的是工业化运作，但是很多人参与。目前，我认为很重要的一点是要利用设计的大环境，我认为现在工业化设计变得很重要，大多数产品都是工业化产品。但是我觉得重点要少倾向于产品，而要放在工厂的运行方式、人们在工厂的工作方式、工厂的选址、工厂的景观、人们在工厂内的着装、人们怎么去工厂工作等方面。在工作的社会中，不光要看到从社会中产出的东西，还要看到人们工作的场所，工作的方式，我对此兴趣更大。

5只猫：您曾经说过"设计是从日常生活中找出美"，您也提到非常喜欢在印度看到的用孔雀羽毛扎成的扫把，这需要设计师保持一颗善于发现美的心，请问如何将这些隐藏在生活中的美找出来，并呈现给大家？

尤尔根·贝：最重要的是不要有偏见，要时刻相信身边的事物是美好的，以及事物在成长中的美。如何将这种美呈现给大家，就是要将这种美解释给大家听。比如，一位作家写一本好书，一名设计师做一个好作品。这是一种语言，这也是一种职业素质。

5只猫：平时您的设计灵感来自于哪里呢？您曾经说过"设

计要像跑马拉松一样坚持"，是否意味着设计师本身除了需要具有相当的设计水准外，还需要一颗勇敢的心来完成设计？

尤尔根·贝：我认为设计师需要在事事都行不通的时候仍然屏住呼吸，坚持下去，这时要想到事情会向下发展的，会慢慢有所进展的，就像一场持久战，而不是短期能够结束的。

5只猫：您的设计作品中我们可以看到大量循环再造的概念，比如绳结凳，比如树干长凳（TreeTrunk Bench），您是在有意这样用环保来做设计么？

尤尔根·贝：不是说环保，而是一种语言，我确信事物都有它的语言，利用现有的事物，就意味着你可以创造一种人们已经听得懂的语言，所以当人们看到的时候，就会有一定的想法。不像一切都是新的，比如如果你对我说中文，我可能会喜欢，但是我听不懂，但是如果利用已有的词汇，我可能能了解个大概。对于我来说，利用已有的东西仅仅是因为可以在已有的语言上实现再造，可以让更多的人读懂。
我是用到了环保，但是我并不是用已有的东西来做一个环保的产品，不是这样的。是你如何得到一种想法，比如说如何使用最少的能源，但并不是说在重新利用，这不是一个环保问题，这是一个"语言"问题。

5只猫：在您看来，设计是从实用出发，还是由美出发？您怎样平衡设计的实用性和视觉美观？

尤尔根·贝：实用，就是在尽可能的范围内从一个处到另一处的实用，就像一个朋友来乡下找你，在一条很长的路上，我向朋友展示各种风景。如果专指高效，我不太同意。我觉得任何事物都是实用的。我不想让我自己只注重效率，有时候需要笨一点，或者笨重一点，这也是我喜欢的特质。
有些东西非常丑但是极其美的，只是因为你很喜欢，或者很爱它们，所以很难说，美是要看具体的环境的。但是我不认为所有的东西都要好看，因为它们会引诱你去做你不会做的事情，所以对我来说很重要的一点做这种产品不是我对其有多大兴趣，我很少做这种东西，只是在我觉得这种很好看的产品对你会产生作用的时候。

5只猫：你到过中国的很多城市，对中国城市的印象是怎样的？

尤尔根·贝：拿香港来说，我很喜欢香港是因为它就像一座科幻城市，城市的建筑风格一半是旧的，一半是新的。街上的行人来自不同的环境、不同的层次。我非常非常喜欢香港，香港就像是一座谜城，相互编织、相互交错、相互开放，这是我很喜欢香港的地方。
1991年的时候我去过北京，那时北京完全是另一个样子，大街上只有自行车，汽车很少见，我很享受那段时间，你可以路过十条街，都是自行车，有点像乡下，我从未见过这样的城市。
到上海我发现我知道很多关于上海的事情，说到喜欢和不喜欢的，我仍喜欢在上海的日常生活，比如切成小块的食物等。

5只猫：您能不能用一句话概括一下，对您来说设计意味着什么？

尤尔根·贝：我想，我之所以喜欢设计是因为事物所拥有的语言，使人们可以做很多事情；我之所以喜欢设计是因为随时随地与人相处，尝试着总是通过事物找到问题的解决方式。设计就是一种制造工具。

5只猫：您认为哪件作品最能代表您？您认为是什么让您能够拥有这么多的奇思妙想？

尤尔根·贝：耳椅最能代表我。
是人与人之间的交往方式，是日常生活给了我创意。

FLOATING PARADISE
飘浮的天空之城

Product Name : Studio Makkink & Bey at the Balancing Barn
Design : Makkink & Bey
Client : Living Architecture
Fotograaf : Studio Makkink & Bey
项目名称：地处平衡谷仓的 Makkink & Bey 设计工作室
设计：Makkink & Bey
客户：利丰建筑公司
摄影师：Makkink & Bey

Dutch design office Studio Makkink & Bey was commissioned to create the interior of MVRDV's Balancing Barn for Living Architecture. Their design was largely inspired by the way that the Barn brings the outside inside and emphasises the presence of both earth and sky.

荷兰设计工作室 Makkink & Bey 受雇为利丰建筑公司（音译，Living Architecture）设计 MVRDV 建筑设计事务所平衡谷仓的室内空间。他们的设计灵感主要来源于谷仓带来的外在内化感，强调大地和天空的现实所在。

The Makkink & Bey scheme traces the transition from earth to sky as one walks from the earthbound entrance of the Barn towards the 'levitating' end of the building, where the floor to ceiling window offers an uninterrupted view of the clouds and sky. Walls and floors are decorated with sampled elements of paintings by Suffolk artists John Constable and Thomas Gainsborough. The classic images gradually evolve into abstract angular colour patches; they become images reminiscent of modern art and they take on a more applied form as wall panelling. The house can accommodate two or eight people equally comfortably, feeling neither empty nor crowded, thanks to customisable furniture and fittings, mostly by Dutch designers. The clear cut Dutch furnishings, primary colours and simple 'butt joinery' used in Studio Makkink & Bey's work all celebrate the elementary design idiom that is characteristic of Dutch design.

当你从谷仓的地面入口走向大楼"漂起"的另一端时，Makkink & Bey 为你勾勒出大地到天空的过渡，地板到天窗间是一幅连续的云天景观。墙壁和地面饰有萨福克艺术家约翰•康斯特堡和托马斯•盖因斯博罗画作的简化元素。经典图像逐渐演化为抽象的棱角分明的彩色小片；变成对现代艺术的暗示，形状更加实用，更适用于墙面镶板。房间可容纳两到八个人，舒适感亦佳，主要由荷兰设计师量身定制的家具和配件让人毫无空旷感或拥塞感。裁减简洁的荷兰陈设，Makkink & Bey 作品使用的主色调和简单的对接木工都响应了充满荷兰设计特色的核心设计风格。

The kitchen acts as a gateway from outside to inside, following the earth theme. On entering one immediately notices the reproduction of a painting by Constable of a horse and carriage that seems to have just walked into the kitchen. Its colours are dispersed over the wall following it round the corner into the corridor. There are six generic sets of cutlery, crockery and glassware and two 'special' sets that were purchased from English companies with a long tradition of producing crockery. Thus two people can enjoy dinner using all the elements of the special tableware, whereas when there are eight visitors, each can have one piece from the special set.

厨房作为内外间的门径，延续了地面的主题。进入厨房，你一眼看见的是康斯特堡画作的复制品，画面上的马和马车似正步入厨房。画面颜色弥漫墙面，到达角落直至走廊。室内陈设有六套普通和两套"特别"餐具、陶器和玻璃器皿，陶器由有着悠久制陶历史的英国公司生产。两个人享受美食的同时可以享受特别餐具带来的所有元素，而当有八名访客时，每个人都可以从特别餐具中选一副使用。

The next room in the building is the Gardener's Room, with a painting of a farmhouse and garden. The form of the fence depicted on one of the painting's fragments and the fences in the countryside surrounding the Balancing Barn served as inspiration for the design of Studio Makkink & Bey's versatile furniture for the house. As well as night stands, the beds all have one plank sticking out to which a reading lamp can be clamped at a preferred height. The night stands double up as stepladders, so that one can reach the upper shelves of the built-in cabinets.

接下来是园丁房，挂有农舍和花园绘画作品。画作一角描绘的围栏形状和平衡谷仓四周的围栏是 Makkink & Bey 工作室设计房间内多功能家具的灵感来源。床和床头柜都支出一个厚木板，可以夹阅读灯，并可以将灯调到理想高度。一对床头柜形成一个活梯，方便够到内嵌式衣柜的顶层。

Next along the corridor, in the Fisher's Room, is a replica of Constable's "The Mill Stream, Willy Lott's House", which shows a mill house with a pond next to it. The water in this painting determines the room's colour scheme. Continuing down the corridor, one arrives at the Pig Room, where the pigs in Gainsborough's painting lend the room its pink appearance. In the adjacent Cloud Room the view is lifted up into the sky. Cloud fragments and grey pixel panels give the room a light tone. Sky and the notion of floating are the key elements of the living room since it hovers over the ground, suspended over the edge of the hill. Studio Makkink & Bey have chosen Gainsborough's most famous painting of a boy dressed in blue to be the signature element of this room, the colour of his clothes echoing and reinforcing the sky theme and informing the colour scheme. Meanwhile, ladies from Constable's "Ladies From The Family Of Mr William Mason Of Colchester" sit quietly alongside the temporary occupants of the Balancing Barn in this light-filled room.

沿着走廊走，接下来是渔夫房，是康斯特堡"米尔小溪和维利洛特之房"的复制品，上面画着一间磨房和房子附近的池塘。画面中的水奠定了房间的基调。继续顺着走廊走，就到了猪猪房，盖因斯博罗画作上的猪为房间带来了粉色的外观。隔壁的云房，景观升到了空中。云的碎片和灰色像素壁板使房间充满了淡淡的色调。天空和漂浮的概念是家具房间的核心元素，因为它在地面上悬浮，在山边飘荡。Makkink & Bey 选取盖因斯博罗的最为出名的蓝衣男孩的画作作为房间的标志元素，男孩身上衣服的颜色与天空的主题相映成趣，重温了主色调。同时，康斯特堡的"考尔彻斯特的威廉·梅森先生一家的淑女"画作中的淑女端坐在浅色房间内平衡谷仓的临时客人身旁。

BLANKETS PRIVACY
毛毯制造的私密时刻

Product Name : WashHouse
Design : STUDIO MAKKINK & BEY
Fotograaf : Gallery Helmrinderknecht
项目名称：洗衣房
设计：MAKKINK & BEY 设计工作室
照片来源：海尔姆新德克耐希特画廊

In Dutch design studio Makkink & Bey (Rianne Makkink & Jurgen Bey)'s world, design is a narrative based process. For their frst solo exhibition at HELMRINDERKNECHT contemporary design gallery, Studio Makkink & Bey has created a site-specifc walk through the threedimensional landscape of a rural village. Mohair woollen blankets hanging on a clothesline give
shape to the space and divide it. Individually woven patterns and lines become a house's exterior walls or refect the imagery of the surrounding landscape. The washing line is draped in between simple wooden posts. Although these posts function as a laundry rack within the space, within
the fctive landscape they become trees.

在荷兰 Makkink & Bey 设计工作室的世界里，设计是一个叙述的过程。在他们的首个"海尔默辛德耐希特"当代设计展览馆的个人展览中，Makkink & Bey 工作室通过乡村的三维景观创造出一种特有的情境。晾衣绳上的马海毛毛毯将空间分割，并给空间构型，而晾衣绳则成了房屋的外墙，展现了周围景观的意象。晾衣绳晾挂在简陋的木杆上。尽管这些木杆在空间内具有洗衣杆的功效，但虚构的景观使其具有了树的表征。

Renowned for continually questioning the role of industrial design in our society, Makkink & Bey approach contemporary questions about the cross-disciplinary feld with innovative solutions. This analytical approach is the basis for their design methodology that not only values beauty, but as well strives towards the continuous

development of alternative means of dealing with space, the environment and our resources. Makkink & Bey's new and visionary work is characterised by its focus
on traditional craftsmanship and the use of environmentally conscious resources.

总是在质疑当今社会工业设计角色的Makkink & Bey工作室通过创新的解决方案来解决跨学科领域的当代问题。这种分析方法是他们设计方法的基石，不仅注重美观，而且不断拓展和探索处理空间、环境和资源的多种途径。Makkink & Bey工作室新鲜而富有远见的作品既有传统技艺的痕迹又有对具体环境特殊资源利用的创新。

The installation, "WashHouse" traces our sensual and experiential perception of private and public spaces. Private space is determined by walls and other extreme demarcations. It is therefore protected and cared for by the individual. Public space is moulded by the responsible cooperation between individuals and groups. It is open, without boundaries and based on the principle of collective participation. Woollen blankets are the design objects through which space is defned, with inner and outer mirroring one another. Like houses, woollen blankets offer human beings protection and shelter.
Taking care of warmth and security, they are roofs over our heads and tuck us comfortably in. Both roof and blanket make private, protected moments possible.

这次的"洗衣房"追溯设计者对私人和公共空间的感官和经验认知。私人空间有墙和其他极端的界限，因此受到个人的

保护和关注。公共空间是个人和集体间诚挚的合作。它很开放,没有界限,而且以集体参与为原则。毛毯是划分空间的设计物品,内外呼应。如同房屋一样,毛毯为人提供保护和遮蔽。温暖而安全的毛毯是我们头顶的屋檐,让我们可以蜗居其中。屋檐和毛毯都可以制造私密而安全的时刻。

Studio Makkink & Bey's woollen blankets can simultaneously stage very private, intimate scenes as well as create vast public spaces. The border between private and public becomes blurred. Mohair is used to represent its softness and permeability. But "WashHouse" reaches as well into another level: by employing clothesline to hang the blankets, the association of cleanliness and freshness arise, for which the individual is responsible in both private and public space.

Makkink & Bey 工作室设计的毛毯在保持舞台私密性的同时提供了广阔的公共空间,私人和公共的界限变得模糊。马海毛用来诠释柔软和渗透力。但"洗衣房"又上升到另一层次:通过利用晾衣绳来悬挂毛毯,干净清新的感觉油然而生,而个人则需要同时保持私人和公共空间的干净和清新。

The woollen blankets were produced at the Textielmuseum's workshop, Textielab, in Tilburg, Netherlands. All blankets are two sided and woven with two different colours. There are a total of fve patterns and ten color combinations.

毛毯由位于荷兰蒂尔堡的纺织博物馆纺织实验室提供。所有毛毯都是双面的,并用两种不同颜色编织而成,共有五种花样,十种颜色组合。

KOKONFURNITURE
Kokon 家具

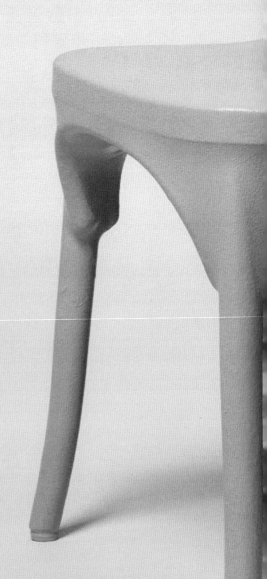

Product Name : Kokonfurniture
Design : Jurgen Bey
Collection Droog Design : Bob Goedewaagen
photo : Bob Goedewaagen/1999
项目名称：Kokon 家具
设计：尤尔根•贝
楚格（Droog）系列设计：鲍勃•哥约德瓦艮
摄影：鲍勃•哥约德瓦艮/1999

These products have been rapped in an elastic synthetic fibre. The material shrinks around a skeleton and forms a smooth elastic skin. The 'skeleton' consists of existing furniture; the elastic skin gives it an entirely new appearance.

上述产品外包有弹性合成纤维。材料在骨架周围萎缩，形成光滑的弹性外壳。"骨架"由现有家具组成，弹性外壳赋予其全新外观。

By cross-breeding and grafting, products and functions of a different nature can merge and develop into new products.

通过材质之间的交叉和嫁接，使不同性质的产品和功能结合在一起，创造出新的产品。

TREE TRUNK BENCH
树干长椅

Product Name : Tree trunk bench
Design : Jurgen Bey
Photographer : Studio Marsel Loermans
photo : Marsel Loermans / 1999
项目名称：树干长椅
设计：尤尔根•贝
照片来源：马瑟尔鲁赫曼斯（音译）工作室
摄影：马尔瑟•洛尔曼斯 /1999

The first benches are chairs-in-a-row. The woods of Oranienbaum are filled with felled trees scattered around which could serve as giant benches. By cross-breeding the trees with a number of different chairs new garden benches are made. An interaction between culture and nature. The tree trunk is the seat, bronze casts of existing backs transform the trunk into a proper piece of furniture.

长椅最初只是一排椅子。奥拉宁鲍姆森林里到处是倒下的树,散落在地上,可以做成很大的长椅。将树木与繁杂多样的椅

EAR-CHAIR

耳椅

Product Name : Ear-chair
Design : Studio Makkink & Bey
Client : Interpolis
Date : 2002
项目名称：耳椅
设计：Makkink & Bey 设计工作室
客户：因特珀里斯（音译）保险公司
时间：2002

The ear-chair has got ears to create privacy and define a space and an arm-rest which forms a small table.
耳椅用耳朵来创造私密性，划出空间和扶手，并形成小桌。

Originally Ear-chair was developed for the reception-room of Interpolis, an indemnity insurance company. There are three different lengths of ears to make different combinations with a various amount of chairs. The ear-chairs form (with at least a couple) the room.
耳椅最初主要是为保险公司 Interpolis（因特珀里斯，音译）的前台区设计的。椅耳的长度有三种。不同数量的椅子有不同的组合，形成这个前台区至少需要一对耳椅。

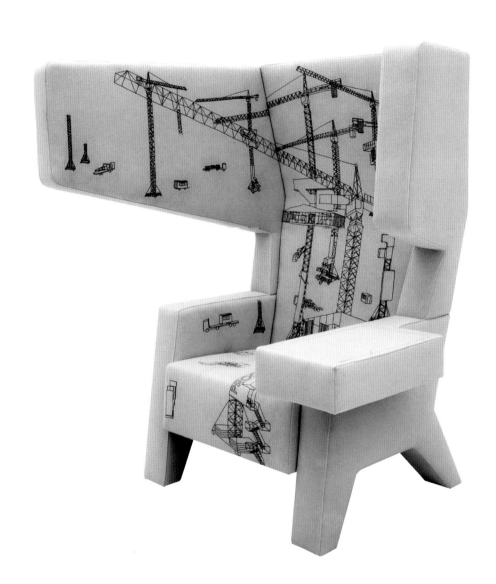

PROOFF

普鲁夫

Product Name : PROOFF
Design : Studio Makkink & Bey/Makkink & Bey
Photographer : Stijn Brakkee
Date : 2009
项目名称：普鲁夫
设计：Makkink & Bey 设计工作室
摄影师：斯提金•布拉基
时间：2009

PROOFF stands for product development for the progressive office. It is a platform for sharing knowledge and exchanging ideas. It gives clients, architects, consultants and designers the freedom to explore different opportunities and come up with unexpected perspectives.
PROOFF（普鲁夫）的全称是进步产品发展办公室。这是一个共享知识和交流想法的平台，为客户、建筑师、咨询师和设计者提供充分发掘各种机会和奇思妙想的自由。

PROOFF offers a range of products designed to help make offices more inspiring places to be. Most of these products were originally intended to solve specific problems for individual clients. For example, how do you create a piece of furniture that allows its user absolute privacy within a huge, crowded public space? The solution is PROOFF's earchair, a product as unique as the brief it sprung from. Because each PROOFF product results from such a process, it is quite unlike any other item of its kind.
普鲁夫提供一系列办公产品,让办公场所富于启迪,发人深思。大多数产品起初都是为个别用户解决具体问题。例如，如何在庞大拥挤的公共场所中打造完全私密的办公空间？解决方案就是普鲁夫耳椅，既独特又简洁的作品。因为每个普鲁夫作品都源于这样的设计，所以铸就了他们的独一无二。

Prooff Work Light
普鲁夫工作灯

Work Light stems from a product range made for Prooff and is the creative foundation from which Studio Makkink & Bey develops new office furniture and products. These pieces generate visual strategies by enclosing and giving space, offering a double meaning or function. The Work lights' beams encircle a small area, gently delineating that space.

It can also be read as an artificial sky, bringing the outside, inside. At the same time it can be put outside to function as a lamppost for office events in the open air.

工作灯为普鲁夫打造的一个产品系列，为 Makkink & Bey 工作室制作新型办公家具和产品奠定了创新基础。这些产品通过封锁和制造空间来增强视觉效果，赋予空间双重含义和功能。工作灯的灯柱环抱出一个狭小的区域，以轻柔的线条描绘出空间感。用户也可以将其理解为一片人造的天空，将外面的世界腾挪进室内。同时，还可以将工作灯放在室外，作公司活动的灯柱用。

Prooff Earchair
普鲁夫耳椅

The Earchairs from Studio Makkink & Bey reveal their possibilities immediately: They are comfortable, have small tables integrated into the armrests and are eminently suitable for defining smaller and larger areas. With the Earchair you create a room in the space available, both visually and acoustically. Variations in upholstery and colour provide moreover the possibility of giving each 'room' its own character.

Makkink & Bey 工作室制作的耳椅直接展现了椅子的各种可能性。耳椅舒适性极佳，扶手部分伸出的小桌很适合划分大小区域。有了耳椅，你就可以在现有空间内分割出一个房间，制造独立的视觉和听觉空间。耳椅不同的装饰和颜色可以为每个房间增添独具特色的亮点。

Jurgen Bey designed the Earchair in 2002 for the reception area of insurer Interpolis in Tilburg.
Not only did the design harmonize perfectly with the ambitions of the principal, it also proved capable of freeing itself from its origin. SV Interieurgroep Rotterdam markets the design under the label PROOFF. The chair is the first and future-oriented design of this range of office and project furnishings for modern businesses.

尤尔根•贝于 2002 年为位于蒂尔堡的因特波利斯（Interpolis）保险公司的前台设计了耳椅。设计不仅完美地契合了委托人

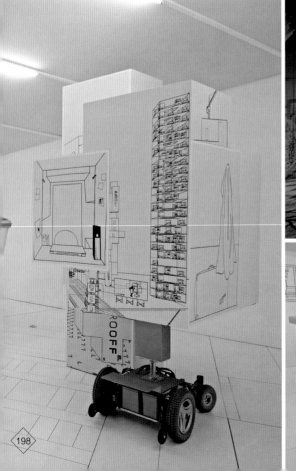

的野心,而且也证明了设计可以从设计的本源解脱出来。鹿特丹 SV 室内设计公司将该项设计贴上普鲁夫的标签,并推向市场。其椅是现代商业办公租赁装饰系列未来设计的首发产品。

Proff Work Sofa
普鲁夫工作沙发

The Work Sofa is a product which takes standard furniture concepts and turns them upside down. In a standard couch, your choice of seating is limited. Here, adapt the Work Sofa to suit your needs, moving its component pieces like a giant puzzle, exploiting any one of a number of possible arrangements

工作沙发吸收了家具设计的标准理念,并将这种理念完全颠覆。一般沙发的坐姿是有限的。但工作沙发却可以让你按需调整坐姿,它的各个部分就像一个大大的拼图,发掘任何一种可能的排列组合。

The Work Sofa's a modular structure is ideal for a number on purposes, including conferencing, collaborative working and (should you require) creating a private space.

工作沙发为会议、合作和 (如果需要) 构造一个私用途提供了理想的单元化结构。

STUDIO MAKKINK & BEY
Makkink & Bey 设计工作室

Product Name : Studio Makkink & Bey
Design : Studio Makkink & Bey
Photographer : Courtesy of Droog
Date : 2009
项目名称：Makkink & Bey 设计工作室
设计：Makkink & Bey 设计工作室
图片来源：楚格（Droog）友情提供
时间：2009

Architecture, furniture and storage all in one. Studio Makkink & Bey took standard sheets of ply wood and with CNC cutting, created a miniature house to stand within larger interiors of offices, studios and lofts. With walls of stool, bench and table parts that easily come out and assemble, the functionality and character of the house can be changed as more or less furniture is used.

集建筑、家具和仓储于一身。Makkink & Bey 工作室利用标准的胶合板和数控切割板，在办公室、工作间和阁楼等宽敞的室内空间里打造出一个迷你房屋。长凳、长椅和桌子的组件易拆卸、易组装，并封闭成墙，使得房屋的功能和特点随着室内家具的多少而产生变化。

A poetic vision for efficient production and material use, House of furniture parts transports flat and can be made to suit different functions, produced locally and customized. Why don't you tell us what yours could be like?

对高效生产和材料应用充满诗意的视角，使得家具的部件可以实现平面运输，并能够满足不同的功能需求，实现了本地生产和客户化。为什么不告诉我们您的家会是什么样子呢？

"If it is a studio, tables and desks will come out. If it is a film house, a projection screen with benches and stools will come out. If it is a children's home, small furniture with cars, animals and a playground will come out," suggests Jurgen Bey.

"如果这是工作室，那么圆桌和书桌都会跳出来。如果是拍摄间，那么长椅、长凳和投影屏幕就会跳出来。如果是婴儿房，那么汽车、动物和游乐场等迷你小家具就会跳出来，"尤尔根·贝说道。

UTRECHT MANIFEST 2009
2009 乌德勒支宣言

Urgent Methods: Design Education Examined
探索课题：检视设计教育

Product Name : Utrecht Manifest 2009
Urgent Methods : Design Education Examined
Design : Studio Makkink & Bey
Photographer : Johannes Schwartz
Date : 2009
项目名称：2009 乌德勒支宣言
探索课题：检视设计教育
设计：Makkink & Bey 设计工作室
摄影师：约翰·施瓦茨
时间：2009

Utrecht Manifest explores Dutch design education in terms of its principles, social ambitions and results. It explores how key figures in Dutch design education hold up specific agendas to their students, under which influences these agendas evolved over time and to which end. The exhibition presents interviews and shows examples of student projects which the tutors and heads of design departments consider to be typical of their vision of the discipline.Urgent Methods presents several clearly distinctive visions of the design discipline by six key figures. Three courses, six tutors and eighteen students.

乌德勒支宣言旨在发掘荷兰设计教育的原则、社会目标和结果。深究荷兰设计教育的核心人物在教育学生时使用的具体大纲，以及这些大纲受到何种影响随着时间的推移不断演变，并走向何方。展览展出了采访的内容，并向外界展示了设计系导师和领导认为在其学科视野内属于典型的学生项目案例。紧急方法体现了六位核心人物对设计学科的真知灼见。三大课程、六大导师和十八名学生。

GASTGASTGEBER
热情东道主

Product Name : Exhibition GastGastgeber - Ruhr 2010
Design : Studio Makkink & Bey
Photographer : Obh Jet Christiaanse
Photographer : Obh Du fotograaf
Date : 2010 年
项目名称："热情东道主"展览——鲁尔区 2010
设计：Makkink & Bey 设计工作室
摄影师：欧伯•杰特•克里斯提亚斯
摄影师：欧伯•杜
时间：2010

For RUHR 2010 Studio Makkink & Bey designed guest rooms in the water tower of the station of Oberhousen, which will open Saterday 1st of May. GastGastgeber will be opened untill October 31th 2010.

针对"鲁尔区 2010"活动，Makkink & Bey 工作室受邀在奥伯豪森中央火车站水塔上设计了酒店客房。活动于 5 月 1 日正式拉开帷幕，并持续到 2010 年 10 月 31 日。

The RUHR area is this year's European Capital of Culture. The GastGastgeber project, a collaboration between Dutch and German cultural institutions, is planning a number of interventions at diverse locations on this occasion. The Dutch are guests in the area, but will act as hosts at the same time: Gast Gastgeber Hans Venhuizen commisioned Studio Makkink & Bey to design a temporary hotel in eight rooms in the old water tower of the train station of Oberhousen Hauptbahnhof.

鲁尔区是 2010 年欧洲的文化之都。"热情东道主"项目是荷兰和德国文化机构的合作项目，计划在此期间穿插一系列以不同场地为背景的项目。荷兰人是鲁尔区的客人，但同时又是东道主：热情好客的东道主。
Hans Venhuizen 委托 Makkink & Bey 在奥伯豪森中央火车站的古旧水塔上设计一座拥有八间房间的临时酒店。

Designing the hotel rooms, Studio Makkink & Bey will perform mostly as a curator. The interior of these rooms will be decorated with a wooden construction with a flip-up bed, so the rooms can serve as public workrooms as well as private bedrooms. An artist will be chosen for each room to exhibit a piece. A guest-artist will then be invited to respond to this exhibit by making a new piece during his stay at the hotel.

Makkink & Bey 酒店房间的设计经验丰富。这些房间的内饰主要采用木结构，放有折叠床，既可公用又可做私人卧房。每间房间会展出一名艺术家的一件作品。艺术家作为客人，会在酒店居住期间，对这些展品做出反馈，并制作新的展品。

For RUHR.2010, European Capital of Culture, Studio Makkink & Bey was asked by Bureau Venhuizen to act as a curator in the GastGastgeber project. The event opens on Saturday the 1st of May in the watertower at Oberhausen station and will run until the 31st of October 2010.

Makkink & Bey 工作室针对鲁尔区 2010——欧洲文化之都项目，受文汇钦当局邀请，主持"热情东道主"项目的设计工作。活动于 5 月 1 日在奥伯豪森中央火车站水塔正式拉开帷幕，并持续到 2010 年 10 月 31 日。

EDUCATION SPACE MUSEUM BOIJMANS VAN BEUNINGEN

教育空间博物馆

Product Name：EDUCATION SPACE MUSEUM
Design：Studio Makkink & Bey
Photographer：Roel van Tour
Date：2008

项目名称：教育空间博物馆
设计：Makkink & Bey 设计工作室
摄影师：罗凡图尔
时间：2008

各部分相互连接，衍生出小的片段，不同的艺术品和参观者以及学生都成了构图的一部分。展览演化出产品和空间，艺术和行为都将成为迷你的展览。

The exhibition spaces become specific spaces, where you can study, welcome people or watch movies.
The study cases are glass desks, where you are surrounded by the art you study, but they also are a place where you can withdraw. The movie case can challenge art academies to make their yearly introduction movies in which the collection of the museum would be the starting point.
展览空间变成实际的空间，你可以在这里学习、迎宾或者观看电影。
学习单元由玻璃书桌组成，周围被艺术环绕，供你研究，但同时你也可以全身而退。电影单元可以挑战艺术类院校，让他们把博物馆的收藏品作为每年介绍影片的起点。

The tables can be used for workshops in which miniature worlds can be made and displayed. Exhibiting, practising and viewing, the education space stays a museum hall, an exhibition by a study collective, witch the objective of new insights and the art of sharing.
博物馆中摆放的桌子可用来制作和展出工作间的迷你世界。
展出、实践和观赏，教育区域在博物馆大厅中，成为集中学习的展区，迸发新的观点，共享艺术成果。

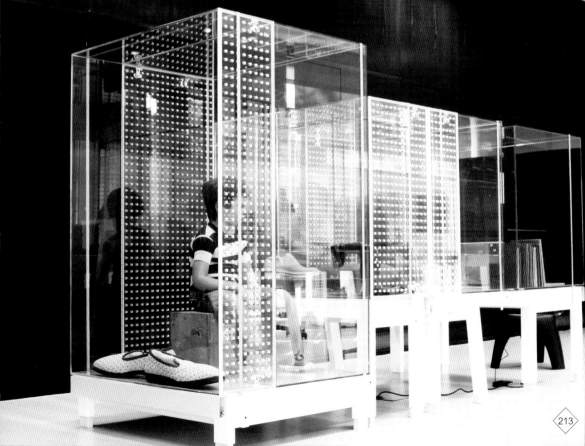

VIJVERSBURG,BRINTA

Product Name : VIJVERSBURG,BRINTA
Design : Studio Makkink & Bey
Photographer: Aly van der Mark
Date : 2008
项目名称：VIJVERSBURG, BRINTA
设计：Makkink & Bey 设计工作室
摄影师：阿里•范德•马克
时间：2008

Studio Makkink & Bey (Architect Rianne Makkink and Designer Jurgen Bey) gets inspired by what is present. Every solution already exists in the world surrounding us. It is only a matter of recognising it and converting it to something useful.
Makkink & Bey 工作室的设计灵感来源于现实世界。每一套解决方案都是我们周围世界的体现。我们仅仅是发现它们，并赋予其价值。

This basic idea was put into practice at the 'Vijversburg VI' estate, by working from the plough lands, that can provide both raw materials for our food as well as construction materials for a place to stay in.
The studio ploughed the land, sowed wheat and built a watch-house on the existing brim of the stork nest.
该项目的基本理念是将"Vijversburg VI"地产变成现实，我们开垦荒地，从土地中汲取食物的原料和居住房屋的建筑材料。工作室开荒、播种，在鹳巢边上建起了观察室。

Straw, the residue of wheat, was stored in crates, which were then piled up to make the house.
我们将稻草和麦子的残余物储存在箱子中，并用箱子堆砌了房子。

The way of building created an effect that is similar to timber framing, but is instead built up out of transportable modules. Just like the other follies on the estate, it is

suitable to repose in for a while.
The view on the pasture carries a new perspective, in the field of grain the letters 'Brinta' can be seen, which closes the cycle: The wheat gives an oath to the well-known breakfast-product with which generations have grown up.
这种建房的方式类似于木框结构的房屋,但取而代之的是便于运输的单元化组件。就像地块上的其他铁物一样,这里也适于小憩。牧场展现的是一片全新的视野,长满稻谷的天地依稀可见"Brinta"的字样,使项目完美收工。用麦子作为原料是对一代代人赖以生存的众所周知的早餐原料的最好讴歌。

SLOW CAR

慢车

Product Name：SLOW CAR
Design：Studio Makkink & Bey
Photographer：Courtesy to Studio Makkink & Bey
Date：2007
项目名称：慢车
设计：Makkink & Bey 设计工作室
摄影师：Makkink & Bey 工作室
时间：2007

Jurgen Bey says, the slow car is part of a series of rethought infrastructures that speculate about the way we make places to work. His ideas attempt to suggest how we might turn the generic into the specific. He has made wall-sized illustrations of massive cranes building modular skyscrapers and themselves being incorporated into the buildings. At their foot are beetle-like cars that form the personal space of this plan.

尤尔根　贝说慢车是基础设施重构想系列的一部分，是对我们打造的工作场所的大胆猜想，他想要化腐朽为神奇。他制作墙面大小的用来建筑模块化摩天大楼的大型起重机图片，并将其融入到建筑中。底部是甲壳虫式的汽车，用来建构计划中的个人空间。

While the idea of the slow car has many obvious and practical effects on road safety, the environment etc, its most interesting potential is social – it is a machine that could transform what Marc Augé would call the non-places of the contemporary world into places more like a campus.

慢车的概念对道路安全和环境造成了很多明显而实际的影响。最有趣的潜在影效是社会性的——它将马克•奥热所说的当代世界的非空间转化为像大学校园的地方。

Think of an airport. It is a sprawling mass of a building, made completely contingent and defined by security and pedestrian mobility. There is no exterior public space at all. There are shopping centres but no public spaces. There is nothing particular, and no hope of it. But imagine an fleet of moving rooms, places you could sit, sleep, keep your luggage, and use to transport you to a park or a church. If all corners of the place were within reach to every passenger then how might that transform the character of these generic places? How much more time might one want to spend there? Where would the civic centre of an airport be, and what would it look like?

试想一下在机场。机场的大楼繁忙嘈杂，完全是随意的，以安全和行人的机动性为基础。根本没有外部公共空间可言。有购物的场所，但没有公共空间。没什么特别的，也不能期待有什么特别的。但想象一下有一个移动的房间，你可以在里面坐着、睡觉、放行李，并带你去停车或去教堂。如果所有的地方都触手可及，那么这些普通的地方将会变成什么样子？一个人还会在那里多花多长时间？机场的市民中心会在那里，而那里又会是什么样子？

The airport is the funnel through which pass the skills, knowledge and talent of any international industry. Imagine if those people could drive themselves to a park, or perhaps into a huge library, teeming with tiny, personalised cars, beetling around like workers in a quarry. This project is an enabling infrastructure of mobile rooms that could make shared spaces in generic places. Beyond the airport, the 40km/h speed limit of the Slow Car slows down the city. Perhaps the campus model could work at this scale. The Slow Car suggests mobility at the scale of the largest city, but not the scale of a country. The city is the Slow Car's field of operation. As a result, it is liberated from aesthetic conventions that dramatise mobility. It is clearly free from aerodynamics, for example. It has another purpose. It is more like a small building, a shelter that allows us to experience public spaces at a much larger scale. It allows us to live at the scale of the city, extends the territory you can call your neighbourhood.

机场是供人们穿梭于国际产业技术、知识和智慧间的通道。想象一下，如果那些人可以自己开车去停车或者出入大型图书馆，或者集合小众，将自己的车子个性化，像采石场的工人那样繁忙往来，那会是什么样子？这个项目就是要将基础设施变为移动的空间，你可以在普通场所分享空间。在机场外，限速40公里/时的慢车可以将城市的节奏慢下来。也许大学的模型可以在这种规模下发挥作用。慢车宣扬在最大的城市规模下实现机动性，但不是以国家为尺度。慢车行动的场所在城市。从美学的惯常中得以释放，渲染灵活的机动。它更像是一座小型建筑，一个让我们在更大的范围内享受公共空间的栖身之所。能够让我们在城市的范围内舒适生存，将邻里的领域延伸开来。

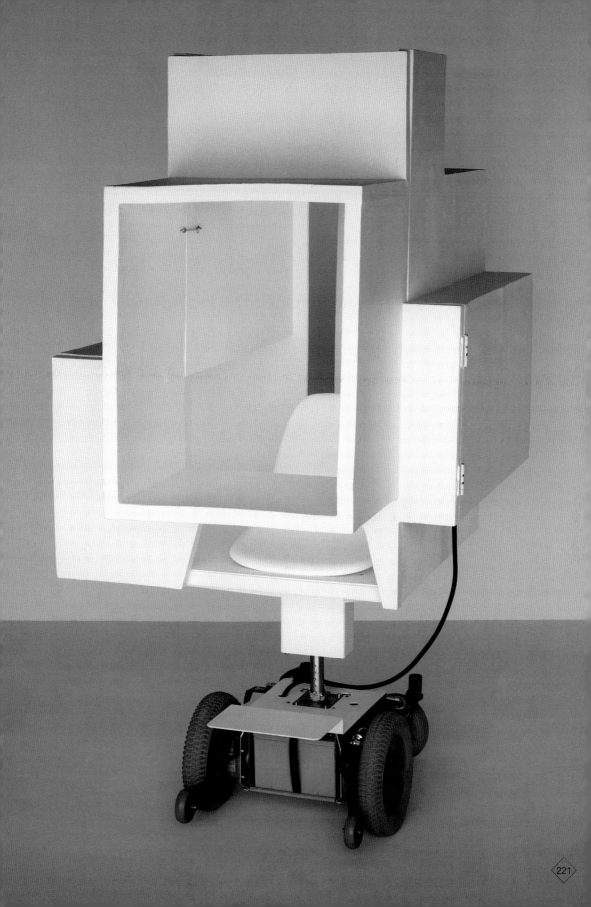

BROKEN FAMILY

破镜重圆

Product Name : Broken family
Design : Jurgen Bey
Photographer: Bob Goedewaagen
Date : 1999
项目名称：破镜重圆
设计：尤尔根·贝
摄影师：鲍伯·歌德瓦艮
时间：1999

Dutch designer Jurgen Bey considers himself a researcher. In his quest to discover the hidden qualities, stories, and emotional value of the objects that surround him, he focuses on their narratives and invests them with new meaning.

荷兰设计师尤尔根·贝认为自己是一名研究员。在发现周围事物的潜质、不为人知的一面和情感价值的同时，他更愿意诠释和赋予事物新的意义。

Back in the late 1990s, he started to trawl his old neighbourhood for furniture, second-hand items, and objects discarded by friends and covered these voluntary cast-offs with a new skin. The result: an eclectic array of functional objects that are intentionally a bit 'off' — works that cultivate, pursue, and cherish a good mistake.

早在 20 世纪 90 年代末，他就开始搜罗邻居的旧家具，朋友不要的二手物品，给这些"旧物换新颜"。结果，看上去有点过时的有用旧物——因为美好的错误而变得有营养、有价值和令人珍惜。

One of his studio's earlier works, the silver-plated Broken Family tea set, hails from this period and assembles mismatched — and often flawed — ceramic pieces into a makeshift, yet elegant patchwork family, united by their metallic sheen. Barely covering the obvious idiosyncrasies, yet thicker than blood, the silver adds a thin veneer of cross-species familiarity. And where the cracks begin to show, they add a fragmented sparkle to the work's reflection on reuse, identity, and reinvention.

工作室的早期作品之一，银盘破损茶具套件就是这一时期的作品，一些相互不匹配、有缺陷的瓷器碎片，经过改新，变成了一套有统一金属光泽的错落有致的茶具。设计师几乎没有掩盖各部分的明显异质，但他们比一家人还要亲，银色在不同的部件上增添了一曾薄薄的外饰。而裂缝出现的地方，为作品增添了残缺的光亮，使它们闪耀着再利用、再定义和再发现的光芒。

LIGHTSHADE-SHADE
燃亮的灯

Product Name : Lightshade-shade
Design : Jurgen Bey
Photographer: Maarten van Houten
Photographer: Studio Makkink & Bey
Date : 1999
项目名称：灯罩•罩
设计：尤尔根•贝
摄影师：马顿•凡•豪顿
图片来源：Makkink & Bey 设计工作室
时间：1999

The lightshade-shades shade the old lamp. They are a new skin for the old, not fashionable, worn-out shade or any other reason not to fit in the new place. The lightshade-shade is a one way mirror that reflects its environment and conceals-reveals its former identity. Turned on, the lamp gives light, and appears in its old garment to give comfort and character.

灯罩•罩用来遮罩老灯，是缺乏时尚感、无论如何无法融入新环境的破旧老灯的新衣。灯罩•罩在某种程度上是一面镜子，它反映了周围的环境，掩盖、揭示了其原有的身份，重新燃亮的老灯在原有装饰下体现出舒适感和一种独特的特质。

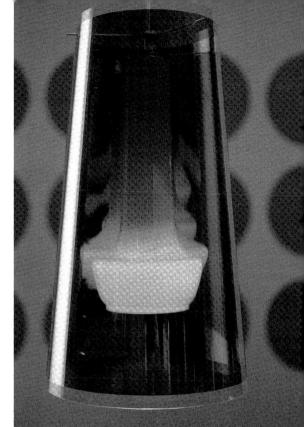

KAdE

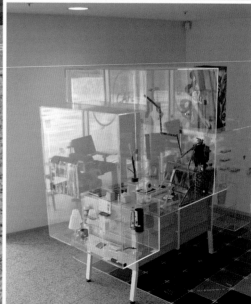

Product Name : KAdE
Design : Studio Makkink & Bey
Photographer: Studio Makkink & Bey
Date : 2008
项目名称：KAdE
设计：Makkink & Bey 设计工作室
图片来源：Makkink & Bey 设计工作室
时间：2008

The KAdE chair is a design object that encourages active sitting. Strong contrasts in colour, material and design between the base and the seat focus attention on the latter, a well-known Eames design. The wooden structure supports the plastic seat wherever balance is required. The design encourages the user to move around in the chair; as Rietveld used to say in defence of his angular designs, "To sit is a verb".

KAdE椅设计产品让人想怎么坐就怎么坐。座椅色彩鲜明，底座和座位间的材料和设计感着重于宣传后来广为人知的埃姆斯（Eames）设计。木质结构随时为塑料座椅提供平衡。而用户可以随意旋转座椅；正如里特菲尔德为其棱角分明的设计辩护时常说的，"坐是一个动词"。

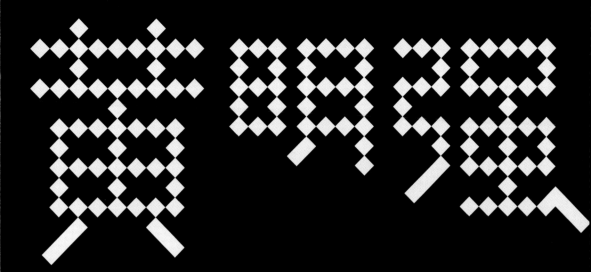

MICKEY HUANG

Paper Down, Mind Up

放下图纸 思绪游弋

MICKEY HUANG
黄明强

1969年出生于中国广东汕头市。深圳市汇能空间装饰设计有限公司的创始人。

黄明强这位来自"设计之都"深圳的中国设计师，不仅运用设计诠释自己对中国文化的充分理解，更通过系列对中国传统文化诠释的设计理论，支撑起对中西方文化碰撞和融合的独特见地，用国际化的手法表现设计的先进特质，视野开阔，使中国设计发出耀眼的光芒。

经过多年实践，中国文化的沉淀和积蓄，在他的设计中运用得出神入化。这种对设计的感悟，得益于个人良好的成长环境以及从业之后，对设计领域孜孜不倦的探索。

潮汕地区是中国的南方城市，这里常年气候湿润，温暖潮湿，独特的地理环境，滋养了这里居住的人，情感丰富，细腻温和，这给了黄明强观察周边事物的敏锐观察力，商业巨子李嘉诚正是潮汕人士，他的成功，成为当地有口皆碑的励志故事，出于祖辈的教育，因此勤奋好学的品质也几乎是黄明强与生俱来。

和每个在求学阶段的设计师无异，为丰富自己的眼界，黄明强多次到贝聿铭设计的苏州博物馆观看和学习。这座建筑是一个保留粉墙、去除灰瓦屋顶的设计，而不是西式的平屋顶或典型的苏州灰瓦飞檐。它突破了中国传统建筑的古典精神，在融合了中国文化精髓的理念之上，融合了西方建筑的现代风格。这种用现代手法表现传统民族文化的优秀设计，给了黄明强极大的启发，从那以后，几乎他的每个作品都能看到这种鲜明的设计风格。

2010年由中国陈设艺术委员会主办的晶麒麟设计大奖，13名国际设计专家评审团分别来自美国、马来西亚、新加坡、意大利、日本、澳大利亚、中国香港、中国台湾等，在全国400多个参赛作品中仅选拔30个优秀的设计作品予以颁发奖项，竞争异常激烈。在这次大奖赛中，黄明强的设计作品《广东伟业陶瓷展厅》荣获奖项。而这个作品，主要采用了北斗七星的理念，通过北斗七星七个点，串联整个空间，这整个的空间概念是围绕着潮州的传统文化根基进行的设计创作，正是对"现代的手法来表现传统的文化"的生动诠释。

黄明强设计的首要原则是空间为人服务，因此室内各空间之间的借景、透景、层层递进和纵深延展，都加以极为细致的考虑；他的灵感从靳埭强的水墨平面设计、敦煌的色彩斑驳的壁画图案、萨顶顶喃喃自语犹似梦呓般的清澈歌声到古琴的溪水流转的袅袅之音，就像和黄明强一同体验这丰富多彩的绚丽的人生。

ENERGY FIELD OF DESIGN
设计的能量场

我们中国的古人创造了许多的文化智慧,"五行"是其中之一。

金、木、水、火、土为阴阳运动变化过程中五种不同量变的存在状态,从阴阳是生化宇宙间一切事物的基础材料这个认识出发,宇宙中多姿多彩的所有事物形式,都是不同量变状态的阴阳相互作用而产生,在不同的阴阳量变对比中,自然会产生出不同的事物形态和运动状态。这就是大地自然的水生木→木生火→火生土→土生金→金生水的循环状态。

中国古代推崇"天人合一",人是天地之间非常非常小的一种生物,其实人很渺小,理解这一点,也就不难理解空间是为人服务的,敬天爱物,然后我们要尊重这个规律,尊重人在空间里的地位,就不会强权于空间,或者独霸于空间,当你尊重空间的时候,空间的每一个角落都尊重你。从整个大的方向是这样的规律。

我们中国人在做中国建筑的时候,包括地理、风水,这些都是很讲究的,室内设计是离不开建筑的,也离不开地貌,我们中国古代人选房子,好的风水以及公认的好的房子的位置应该是群山环抱,房子前面要有水,讲究这种明堂,后面要有山,也就是房子要有山有水。水寓意财富,山寓意靠山,人离不开水,离不开空气,设计要巧用穿堂风,水也要巧用,上善至水。这样才形成这个房子的能量场。其实这个场,在这里是指"风水"。

室内的空间布局,这是设计师要首先考虑的,之后在设计里调整和运用多重元素:新颖的环保材料,雅致单纯且含蓄内敛的色彩,年代久远依然色泽艳丽的陈设品,过滤了杂音的纯美动听的音乐,常常被人忽略了的若隐若现的气味,明亮或灰暗的色、淡淡缭绕的香、浓烈或刺鼻的味,我们的眼睛所能看到的感官,人的身体……空间不是一个冰冷的地方,它是可以触摸的,如果我们走进一个空间,什么都不能动,那实际上这个空间已经死了,你要和空间互动。

中国古代强调:人少,房子不要太大,因为这个自然形成一个场。作为现代人的我们常常说:空间因人而鲜活。空间是为人服务的，那么你的空间是为谁服务的？设计，绝对不能忘记了初衷：为人服务。那么为人服务，空间因人而鲜活，你设计了一个空间，但是人不愿意在里面呆着，那你设计的这个空间还有意义吗？

空间的灵魂是什么？
是文化。
文化，就要找到它的根，就是能量场的问题。
空间装修完工的时候，
人愿意在这里呆着，
你喜欢和这个空间互动。
空间也因为你的喜欢而更加回报于你。

能量场，其实很多人都没有办法感受，认为"眼见为实"。

眼睛所见的只是20%，你以为你看见了，其实80%你是看不到的。很多人会很奇怪。那我可以问一下：信息波，电波，你看得到吗？空中飘浮着许多的细菌，你看得到吗？是不是都看不到！

灵魂你看得到吗？有人会反驳：没有灵魂啊，这个世界上没有灵魂。

你如果想念一个人，这个人也想念你 你拿起电话刚要打的时候，刚好对方也在给你打电话：怎么这么巧，我也想你啊，我正在给你打电话！这就是我们常说的"心有灵犀"，其实这就是相互之间是存在着信息波的。
这就是信息波，你看不到。

我们中国医书上写，人的身上有穴位，有气脉，在1848年以前，世界医学史不承认中国的中医：你们中国中医是胡说！穴位在哪儿？我测不到。但是，1970年，美国科学家研究出了一套仪器，真的能测出人体的穴位以及经脉，科学家测出了中医医学的能量场。

我们传统的道家、佛学，在说到修为方面，在说到禅道入定的时候，可以看到身体上的七层气场，包围了整个人体。身体越好的人的气场越多：身体弱的人包围他的气场是一层、二层、三层；病人也有；死亡的人就没有气场了。以上这些就叫能量场，是眼睛看不到的。

所以，等到西方的科学家用仪器测出能量场的时候，他们发出赞叹：哎！中国人讲得对啊！

《孙子兵法》，目前许多的大国把它当成一本圣经，还有我们的《道德经》、《易经》，为什么被国外的那些科学家、社会学家、经济学家、管理学家、医学家研读？是因为里面蕴含着许多的信息密码。医学、计算机，甚至设计，都能从中找到相应的信息密码，比如我们说设计，其实道理是万物相通的，有时我们说：设计要密不透风疏可跑马，和阴阳，和国画，知白守黑，道理是一样的，画国画的时候，你要惜墨成金。

能量场和空间设计的关系，其实是一样的。

其实这种空间的能量场,实质是设计师应用综合知识的能力场，把一个空间表现得精准传神，这其实是设计师对空间能量的体现。

INTERVIEW: MICKEY HUANG

黄明强访谈

5只猫：我们知道：在设计界，您以强调设计和文化的关联而著称。中国传统文化源远流长，比如说儒家之中庸合一，道家之玄虚飘逸，佛教之禅中沉寂，但是现代设计由西方传播而来，不可避免的这些传统文化受到很大冲击，现代的设计，或者说西方的设计思潮，如何在商业设计的过程中，诠释中国传统文化的精髓？

黄明强：西方的设计风潮，蜂拥涌入中国，这些西方文化的大量涌入，抢占了中国的一些商机，但同时也带动了中国的设计。

一方面把非常先进的科学技术和设计理念带到中国 另一方面把中国我们老祖宗这些东方智慧再一次地深刻挖掘，并提炼出来，让中国的文化有上升空间，从这方面讲，起到了好的作用。

我们中国近现代是一个苦难的过程，从清朝末年到改革开放之前的这段时期，经历了战乱和文革，我们的文化已经遭到严重破坏，再次挖掘文化的历程也比较晚，但通过这三十年的经济高速发展和外来文化的冲击，让我们的中国再一次回归文化正轨。

5只猫：最近几年，西方建筑师抢滩我们中国本土建筑设计，已经形成一种现象：我们的包容性很强，我们有小蛮腰的设计，库哈斯的央视新楼，还有2008年的鸟巢、水立方，国家大剧院等，但是却很难得一见我们中国建筑师类似这种经典的广受传播的建筑设计，您觉得这是什么原因？究竟是我们中国设计师和西方设计师确实存在设计实力方面的差距，还是由于决策人审美品位的偏离和缺失，或者还有其他的什么原因，作为设计师，您怎么看这个现象？

黄明强：这其实是一个很大的议题，其中的原因大概有两点：
第一，高层领导人素质和决策的问题。作为领导人，他要从决策人的角度，来寻找一种平衡。
虽然，站在设计师的角度，有人要骂：这是什么鬼设计？但是决策层要平衡大众的审美，再加上自身素质及一些其他因素的影响，比较少尊重设计师的设计理念。

另一点，从一个中国设计师的角度来看，不得不承认：我们和国外设计师的设计水平还是有差距的。

中国改革开放至今，才这么短暂的时间，真正培养一个有素质的设计师，是需要从小培养，从家庭、环境、教育氛围以及教育体系等多方面培养起来的，所以，这些，是促成我们中国设计师先天不足的条件。

西方的设计教育是启发性教育体系，不是填鸭式的，老师说的就不一定是对的，而是启发式的，让你开动你的脑筋，启迪你自己的智慧，来发掘设计的源泉和设计的理念，这和我们国家的教育体系就不一样，所以造成中国的设计师和国外的设计师设计实力方面有一定的差距，这是根源。

还有受国家经济实力的影响，我们暂时还没有办法像西方那样，深入地研究性地做些课题。一个小小的建筑，一个小小的设计，西方设计师可能用好几年，但是我们中国目前的设计状况不是研究课题式的，它是高速，或者可以说是急速行进的过程中迸发出来的，所以，必然使我们的设计有些浮躁，也有些浅薄，没法做到西方设计的那种内涵。

我很积极地看待西方设计的冲击，俗话说：落后就被挨打，最关键的是你挨打的时候，你学到了什么？你要学什么，不要被动挨打，而要在被打的过程中崛起。其实做设计和做人的道理是一样的。

出现这样抢滩中国建筑的现象，其实也是一件好事，在争鸣和竞争之中，我们学到了更深层次的专业知识，可以快速提升我们的素质。

5只猫：日本的设计，发展得也很快，他们在借鉴和发扬西方设计文化的同时，很好地保留了本民族的特色。我们中国设计师从中吸取什么？

黄明强：日本的本民族文化传承和地域之间有很大的关联。

日本从地理位置来讲，是一个很小的小岛，所以，这个民族从孩子很小的时候开始，就在宣扬危机意识。日本民族始终强调自己国家的资源匮乏，而我们中国总强调自己地大物博。这导致我们产生一种心理暗示："浪费是没有问题的"。我们危机意识非常不够。

日本在学习西方先进文化的过程中，也走过我们中国现阶段的这种弯路。但是他们很快就恢复过来，因为有了这种危机意识，他们时时刻刻都在提醒自己，走偏了路，很快调整回过来，是这种潜在的"质"，决定了他们对自我的把握，这一点，是很值得我们去借鉴和学习的。

当然，我还是很积极地看待目前我们中国设计师和设计行业的状况，现在包括媒体，都在呼吁我们要注重危机意识。

5只猫：我们再讨论下日本设计。日本设计在体现禅意方面较吸引人，这和咱们中国文化也有一些共同之处，最近两三年，我们中国设计师也有类似的作品，但是，在样板房设计这种类型，大多模仿、抄袭所谓的奢华，较少看到那种很空灵、很宁静、很能体现我们中国传统文化的作品。

黄明强：这和文化、经济发展是无法脱离的。

当日本人把某一种物料、物体或材质运到本国的时候，追求物超所值。他还强调可持续的 而不是把一大堆的垃圾往自己的国家搬。

日本把我们中国唐宋时期最好的文化保留了以完整保留，并发扬光大，得以很好的传承。这对设计文化也产生影响起到非常大的作用。

我们中国经历了经济的发展，文化的大变革，继承就变成了半断层的。但是我不担心目前这种文化的断层，我最近在学老子的《道德经》，也看一些中医的书，我感觉真正的文化精髓还是在中华文明。我们现代人应做的是不断地去挖掘，把文化的精髓继承、应用、传播。

我相信我们中国人有智慧的人是很多的，但是这些有智慧的人可能不会浮在表面，修为到一定程度，他逐渐会沉静下来，而当这些人崛起的时候，释放的能量将是不可估量的。

其实，不单纯只是我们中国人反复强调我们的中国风中国文化，西方设计师也在追寻中国风。尤其是最近几年，西方设计师发现，他们的设计风格，已经很难代表设计界的一种倾向和趋势；意识到中国的高速发展，中国的文化，可以说全世界的文明只有中国仍然继承民族文化的脉络持续进步，这是非常难得的，也是我们作为中国人非常自豪的一点。

所以说如果我们文化文脉的根能够保留下来，是我们中国对人类的一个贡献。

5只猫：我们在吸收先进的西方文化的同时，也要保持独具中国风的本民族特色，正如其实每个国家也都努力在探索自己国家的设计风格那般：日本形成日本风，印度形成印度风，中国形成中国风。好的，那么延伸到深圳在2008年被联合国科教文组织评为全球的6个设计之都中的一个，两年多过去了，您作为深圳设计之都的设计师代表，对深圳的设计环境有着怎样切身的感受？

黄明强：其实"设计之都"这个称号是非常令人鼓舞的，但是如果真正的要成为"设计之都"，那就不能只靠设计师单方面的力量。

深圳被评为设计之都 是因为平面设计呼声很高，扛了这面大旗。我们不是说它不好，它是和建筑、室内、服装、珠宝等息息相关，甚至它应是全民的，让大家觉得每个人都可以设计。在国外其实很开放，设计的人不一定是设计师，哪怕是一个打工族，或者是一个高级白领，他们都可以自己设计，DIY。

如果一座城市，总是在强调设计是那一帮设计师干的活儿，已经偏离了设计之都的本意。

政府也要引导，让这座城市里所有的人参与设计，要鼓动他们的参与感和热忱，全民皆设计，当全民参与，这种互动，会把设计推向高潮，推向顶峰，逐渐让设计成为一种习惯，创意也就越来越丰富。市民广泛参与，政府积极支持，设计师努力进步，这样才能真正名副其实地叫作"设计之都"。

5只猫：设计之都也有您的一份力量呢，您的设计作品《广东伟业陶瓷展厅设计》，荣获2010中国陈设艺术委员会主办的晶麒麟设计大奖赛的入围奖，这个奖项的特殊性在于：首先它拥有在设计行业内专业实力雄厚的包含美国、马来西亚、

新加坡、意大利、日本、澳大利亚、中国香港、中国台湾等13名国际设计专家评审团，其次，在全国近400个参赛作品中仅选拔出30个优秀的设计作品予以颁发奖项，竞争异常激烈。您的这个展厅设计，我能感受到强烈的中国传统文化的面貌，请您谈谈这个设计作品。

黄明强：解释一下：这一次很多的设计作品都是很棒的，有时候获奖也是机缘。

我的每件作品我都去寻找它的文脉，它的根基，然后尽量去挖掘，去表现。

这个展厅，从前期构思、平面规划时就加以重点表现。因为它本身是企业内部的美术馆，企业是做陶瓷的，他们收藏了很多的工艺品：陶瓷、木雕、潮绣、字画等，展示在设计中。它是工艺美术的荟萃，非常复杂，不像是一个家居环境：营造氛围就可以了。

当时为他们设计的时候，主要采用了北斗七星的理念，通过北斗七星七个点，串联整个空间，这整个的空间概念是围绕着潮州的传统文化文化根基进行的设计创作。

5只猫：您这个"文化河流"图片的抽象造型，让人感到很震撼，您是用了很现代的手法，表现了很传统的文化，我对表现手法很感兴趣：怎么用现代的手法来表现传统的文化呢？您是怎么找到这种现代手法的？

黄明强：造型其实我也是借鉴了国外的一些设计。潮州的陶瓷文化其实很精彩，只是挖掘力度不够，不像景德镇那样引人瞩目，通过这个陶瓷展厅的设计，我看了许多的历史文献，研究历史资料，发现潮州的陶瓷不愧为"南国瓷都"，从新石器时代，就开始有陶片出现了，演变到现在，所以我就在这个展厅里面设计了一条文化的脉络，叫作"文化河流"，寓意文化源脉的源远流长，通过抽象造型的河流，用根脉相连的形式，把潮州从古至今的陶瓷发展进程得以表现。

我认为我们的设计是为空间服务，是为现代人服务的而不是为古代人服务的，要继承和发扬，所以把最好的一面展示出来，应找到一个亮点，把现代人需要的空间和氛围营造出来。

5只猫：就像张艺谋在2008奥运会开幕式上，运用非常现代的手法表现中国的传统文化：徐徐铺陈的画面、水墨、击缶、和为贵等情节，我觉得您这个设计作品也是这样，用玻璃材料、很现代的空间处理的手法去体现文化河流的这种文化脉络，我很想知道，在您的下一个设计作品中，是不是也是运用现代的手法来表现我们传统的中国文化？

黄明强：是的，现代的设计手法，也可以说是顺应时下现代设计的风潮。

还是那句话：设计是为人服务的，它要具备时代性。设计作品也要具有时代性。为什么古代的那些东西，我们现代的人会很喜欢？是因为它有一个很鲜明的文化特征，我们作为现代设计师，在进行设计的时候，肯定要带有时代的特征，而不能直接搬用古代的。当然有些空间也是需要部分借用的，注意，那是它所需要的。更多空间是需要继承和发扬的，要不然设计的风潮就会停止了。

5只猫：稻盛和夫是日本的经营之神，他有一个观点是"敬天爱人"，我觉得通过您刚才讲的包括你设计的空间，贯穿着"敬天爱人"的理念。您的作品中，特别是融入自然，使得空间非常有层次，我希望在处理空间的延伸和层次感方面您能多谈一点。

黄明强：敬天爱人，这句话很有现代意义。空间所要做的就是尊重人，空间因人在里面而鲜活，那肯定要尊重人，是为人服务的，要适合人的需求，人也不可能脱离自然，要跟自然互动，那没有自然怎么办？要创造一个自然。

同时设计是一个不断的感悟和积累的过程，我做设计21年，这么多年，有着较多的体悟。像我们中国人写书法：读帖、临帖、写贴，读读写写，学习再学习，因为这样的人生阅历，才有自己的体验，它是需要时间的。同样是一个设计项目，有的西方设计师宁肯花费三年五年其至一生来进行设计，我们中国设计师如果很浮躁你能做好吗？知此而后有定，定而后能静，静而后能安，安而后能虑，虑而后能得，你得安定下来，你的心静啊，静的时候又会思索，思索的时候会出创意。如果今天我们总是：哇！赶快！赶紧！如果未经沉淀，那么你的智慧体现在哪里？

空间的层次感也是一样，比如平面布局，图纸往那一放，然后我就想如果我坐在这个房间，从这个角落看出去，会是怎么样的呢？从另一个角落看出去又是什么样的呢？你得想。你要体验，你要思绪游弋，毕竟空间是通透且无限延展的。

5只猫：您刚才讲到了空间尊重人，就是要尊重业主的感受，您设计环境，如何从身心健康角度为业主着想？

黄明强：我最近在学养生中医。你会发现许多道理真的是一脉相承！业主发现空间有问题来找你这个设计师，这时候你是一个医生，你要医治这个空间。说到医治，你要先把好脉，并且理解业主的要求，找到问题点，然后去解决这个问题点，最后就下药，材料很多，我也有很多的配方：墙纸10克，灯具11克，地板20克，窗帘3克……你就要拿捏这个尺度，然后，加水，多少水，然后煮多久，最后这药终于熬出来了，你要趁热喝，这样才能药到病除。

ZEN TRANQUILITY SIMPLICITY
禅·宁静·减

设计单位：深圳市汇能空间装饰设计有限公司
设 计 师：黄明强
项目名称：兰若茶舍
位　　置：深圳市科苑南路海阔天空雅居商铺
建筑面积：230m²
主要材料：编织纹墙纸、竹子、青砖、素水泥

THE BIG DIPPER

北斗七星

项目名称：广东伟业陶瓷有限公司美术馆
设计师：黄明强

本项目通过星象概念——北斗七星作为设计主题，运用现代设计手法对整个空间进行构建，将美术馆的各个展厅，分解为独立的个体，同时又让各个个体建立不可分割的联系，浑然天成，使各展厅在展示其局部的变化关系之时，更彰显整体的大气磅礴，淋漓尽致地演绎了七颗星组成的星阵，即"北斗七星"。

美术馆中展示着以陶瓷为主的传统的潮州工艺品，通过这些潮州工艺品，让参观者直观感受浓郁的潮州文化。木雕、潮绣、石雕、陶瓷、泥塑等和传统的潮州民居空间融合在一起，展示了潮州工艺和特点鲜明的地域文化，散发着民族瑰宝的光芒。

"南国瓷都"——潮州，早于石器时代，已经出现陶瓷了，如何表现这源远流长的历史文化脉络？设计者巧妙地用文化长河将其诠释；河流上是各年代历史时期出现的重大事件，这条经过现代艺术手法处理的抽象河流，充分表达了潮州源远流长的历史文化。

整个美术馆，充满了无数具有中国韵味的细节：经典云纹镂空的雕花花纹，水墨纹理的黑色大理石地板，砖块花纹的墙面，中式屏风等，将中国陶瓷艺术丰富多彩的艺术魅力融于设计手法的现代空间，出神入化地诠释了设计的主旨。

SCROLL PAINTING
画卷

项目名称： 长城集团深圳运营中心艺术馆
设计师： 黄明强

作为长城集团全国市场重要的启动枢纽，深圳运营中心艺术馆有着非常重要的特殊地位。它是长城集团对外的窗口，面向全国，辐射世界。同时为了响应长城集团以陶瓷为载体，传播陶瓷艺术价值的理念，艺术馆处处凸显着长城陶瓷的文化性、品牌性。

艺术馆设计分为长城企业文化展示区、大师成长之路展示区、大师作品展示区三大重点区域，以抽象的雕塑画卷的形式，来展示长城集团企业的核心产品。最引起人瞩目的是蕴含着科技含量、艺术价值和人文思索的四大产品展示区：一、奥运体育瓷——水立方、鸟巢系列；二、废瓷利用——掌握了废弃陶瓷回收、加工并应用于陈设艺术陶瓷生产的核心技术，提高了废弃陶瓷利用率，使项目更加环保；三、长城创意瓷——传承泥与火的灵魂，让中国的陶瓷更加瑰丽。四、潮州陶瓷历史——让顾客通过空间造型来认识及感知长城的品牌形象。四个核心产品区域陈列着甄选的精美陶瓷，给艺术馆增添了无穷的魅力。

作为长城集团展厅三大重点区域的亮点部分，大师成长之路展示区可谓寓意深刻，它为我们揭示了大师成功背后的辛酸：锯齿状的地台、破碎的瓷片以及横在其中的枕木表达了成为大师的艰辛和毅力，层次残损破碎的陶瓷，犹如一幅画卷。生动的画面告诉我们：任何人的成功都不是一蹴而就的，所有的大师都是经历过无数次的失败、艰辛的沥练，才铸就了今日的辉煌。

整个艺术馆，处处显现着对比：动与静、曲与直、古典与现代、科技与经典，这是一种诠释陶瓷艺术的新语言，提炼陶瓷艺术发展的深层文化，唤醒人们对美好的珍惜，对美好的向往。

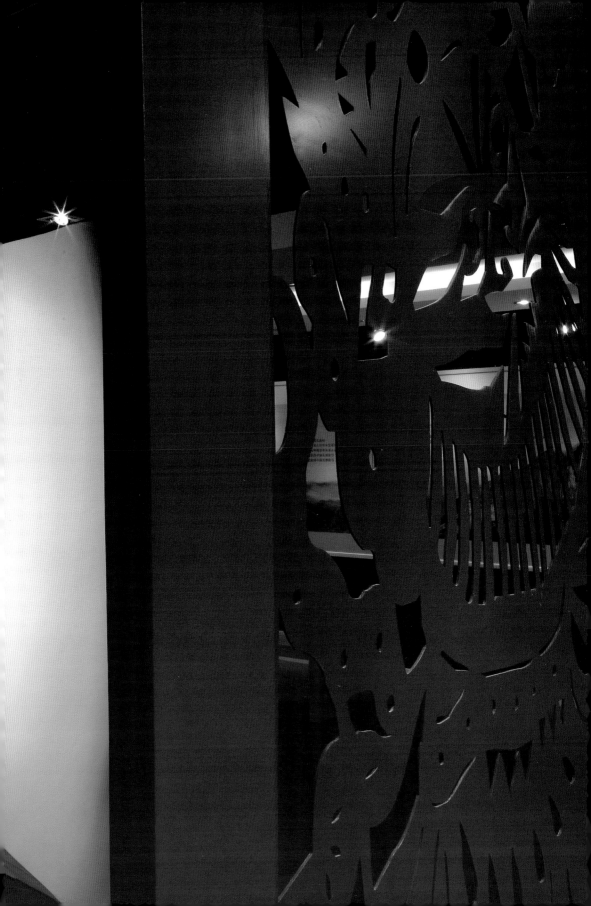

GRASS COLORFUL

芳草缤纷

文 / 燃燃

从高田贤三求学若渴周游世界漂泊大海的那张船票开始，中国、荷兰、西班牙、法国、韩国、日本6个国家的设计师，散发着各自国家民族文化特质，分别讲述了设计作品背后的故事。这些故事比设计作品的本身更动人。这些故事使我们的内心温暖，使我们说话时言语温柔，使我们因设计的美好而更加热爱生活。当我电话采访荷兰设计师JURGEN BEY的时候，从阿姆斯特丹传来的他的声音仍然激情飞扬。JURGEN BEY的思维跳跃，正如他的设计观点：我们就像是受好奇心驱使或肩负使命的探险家，周游世界，不停地发现、询问和关联。我们带回故事，并通过设计诉说这些故事，因为设计就是我们的语言。

那么究竟什么才是设计呢？

我曾看过汶川地震一周年纪念时，广州美术学院师生共同设计的"灾难设计"：小小的拉环扣在担架或者木板上，根据力学原理，它发挥的作用是普通战士扛着木板门奔跑的重量的4倍，也就是说它可以节约减少4个人的工作效率，这在当时人力物力都十分匮乏的情况下，是多么的难能可贵；身为设计师，是不是我们只会"命题作文"：设计豪华的别墅，设计奢华的会所，设计风尚的样板房？我们是否具备设计身边点滴事物的能力？我们身边的家人朋友是否因为你是一个设计师而受惠受益？近两年来，"低碳设计"成为许多设计师热衷讨论的热点，是谁在倡导低碳？是谁在真正受"低碳"之益？又有多少设计师真正践行了"低碳"的设计？低碳究竟是一件华丽的外衣，还是一种隐匿的谎言？

设计是一种感觉你会否注意到其中细如发丝的微小眼神，当你发现身边的同事坐在阳光里，眼前是刺目的光，这时取来一块纸板为她遮挡阳光，发自内心的、顺带而为的行为，却带给人内心温暖，这是不是设计；一个单身的女孩，在寒冷的圣诞夜无处可去，刚刚结识的著名的摇滚音乐歌手，就在当街的大排档为她高歌成名的摇滚歌曲，这是不是设计了温情。

我们居住的生活空间需要设计师妙手回春，我们的心灵更需要每一个充满情谊的朋友设计温情，设计友谊，设计未来。

这就是设计。

设计就在你我的身边。她从未远离，也不会远离。她一直都在，在宁静的黎明和思绪奔腾的夜晚，一直存在。她在凝视着值得珍惜的一切。

葳蕤葱葱的野草，高高地迎风飘舞，墨绿色的枝叶在皎洁的月光下熠熠闪亮。这里不是法国画家卢梭笔下的《赤道地带的原始森林》，这里是春色恣肆流淌的奇异世界，芳草缤纷，这里是设计的天堂。在这里，夜空中，星星闪耀点点银辉；草原上，绽放艳阳的光芒；黑色的大地，生长着万物萌生的无穷的力量。